Lecture Notes in Statistics

Edited by J. Berger, S. Fienberg, J. Gani, K. Krickeberg, and B. Singer

52

Prem K. Goel
T. Ramalingam

The Matching Methodology: Some Statistical Properties

Springer-Verlag
New York Berlin Heidelberg London Paris Tokyo

Authors

Prem K. Goel
Department of Statistics, The Ohio State University
Columbus, OH 43210-1247, USA

Thirugnanasambandam Ramalingam
Department of Mathematical Sciences, Northern Illinois University,
DeKalb, IL 60115, USA

Mathematics Subject Classification Codes (1980): 62C07, 62P99, 62HXX

ISBN 0-387-96970-5 Springer-Verlag New York Heidelberg Berlin
ISBN 3-540-96970-5 Springer-Verlag Berlin Heidelberg New York

Printed in Germany

Printing and binding: Druckhaus Beltz, Hemsbach/Bergstr.
2847/3140-543210

Preface

Incomplete-data problems arise naturally in many instances of statistical practice. One class of incomplete-data problems, which is relatively not well understood by statisticians, is that of merging micro-data files. Many Federal agencies use the methodology of file-merging to create comprehensive files from multiple but incomplete sources of data. The main objective of this endeavor is to perform statistical analyses on the synthetic data set generated by file-merging. In general, these analyses cannot be performed by analyzing the incomplete data sets separately. The validity and the efficacy of the file-merging methodology can be assessed by means of statistical models underlying the mechanisms which may generate the incomplete files. However, a completely satisfactory and unified theory of file-merging has not yet been developed. This monograph is only a minor attempt to fill this void for unifying known models. Here, we review the optimal properties of some known matching strategies and derive new results thereof. However, a great number of unsolved problems still need the attention of very many researchers. One main problem still to be resolved is the development of appropriate inference methodology from merged files if one insists on using file merging methodology. If this monograph succeeds in attracting just a few more mathematical statisticians to work on this class of problems, then we will feel that our efforts have been successful.

We are grateful to Steve Fienberg for inviting us to initiate the monograph and for continued support and encouragement during various stages of the monograph. Thanks are due to Mr. Panickos Palettas for his help in programming the graphical displays appearing in the appendix. We also thank the Department of Statistics, The Ohio State University for providing continued support for this research project and for the manuscript writing. Ms. Myrtle Pfouts did an excellent job of producing the manuscript on her word processor, and she deserves a big 'thank you' from both of us. Finally, we would like to thank the National Science Foundation and the Air Force Office of Scientific Research for their financial support at various stages of the research project under grant # DMS-8400687 and contract # AFOSR-84-0162 .

September 1988
Columbus, Ohio

Prem K. Goel
T. Ramalingam

Table of Contents

Page

Preface i

CHAPTER 1. INTRODUCTION AND SUMMARY 1
1.1 A Dichotomy of Matching Problems 2
1.2 A Paradigm 4
1.3 A General Set-up for Statistical Matching 6
1.4 The Statistical Matching Methodology 7
1.5 Constrained and Unconstrained Matching 8
1.6 Reliability of Synthetic Files 11
1.7 Summary 13

CHAPTER 2. MERGING FILES OF DATA ON SAME INDIVIDUALS 14
2.1 A General Model 14
2.2 Notations 15
2.3 Model-based Matching Strategies 16
2.4 Repairing a Broken Random Sample 18
2.5 Reliability of Matching Strategies for Bivariate Data 24
2.6 An Optimality Property of the Matching Strategy φ^* 25
2.7 Monotonicity of $E(N(\varphi^*))$ with respect to Dependence Parameters 35
2.8 Some Properties of Approximate Matching Strategies 39
2.9 Poisson Convergence of $N(\varphi^*)$ 47
2.10 Matching variables with independent errors model 68

CHAPTER 3. MERGING FILES OF DATA ON SIMILAR INDIVIDUALS 70
3.1 Kadane's Matching Strategies for Multivariate Normal Models .. 70
3.2 Alternatives to Statistical Matching Under Conditional Independence 79
3.3 An Empirical Evaluation of Certain Matching Strategies 86

APPENDIX A 94

APPENDIX B 98

REFERENCES 144

Author Index 149

Subject Index 151

List of Tables

Table 2.1 Expected Average Number of ε -correct Matchings (ε = .01)........ 94

Table 2.2 Expected Average Number of ε -correct Matchings (ε = 0.01)...... 94

Table 2.3 Expected Average Number of ε -correct Matchings (ε = 0.05)...... 95

Table 2.4 Expected Average Number of ε -correct Matchings (ε = 0.1)........ 95

Table 2.5 Expected Average Number of ε -correct Matchings (ε = 0.3)........ 96

Table 2.6 Expected Average Number of ε -correct Matchings (ε = 0.5)........ 96

Table 2.7 Expected Average Number of ε -correct Matchings (ε = 0.75....... 97

Table 2.8 Expected Average Number of ε -correct Matchings (ε = 1.0)........ 97

List of Figures

1.1 Statistical Matching of File 1 and File 2 with information on 6

EDF For Correlation Estimates

3.1 n=10, rho(xz) = 0.00, rho(yz) = 0.10, rho(xy) = 0.00 99

3.2 n=25, rho(xz) = 0.00, rho(yz) = 0.10, rho(xy) = 0.00........... 100

3.3 n=50, rho(xz) = 0.00, rho(yz) = 0.10, rho(xy) = 0.00........... 101

3.4 n=10, rho(xz) = 0.92, rho(yz) = 0.65, rho(xy) = 0.60........... 102

3.5 n=25, rho(xz) = 0.92, rho(yz) = 0.65, rho(xy) = 0.60........... 103

3.6 n=50, rho(xz) = 0.92, rho(yz) = 0.65, rho(xy) = 0.60........... 104

3.7 n=10, rho(xz) = 0.93, rho(yz) = 0.75, rho(xy) = 0.70........... 105

3.8 n=25, rho(xz) = 0.93, rho(yz) = 0.75, rho(xy) = 0.70........... 106

3.9 n=50, rho(xz) = 0.93, rho(yz) = 0.75, rho(xy) = 0.70........... 107

3.10 n=10, rho(xz) = 0.94, rho(yz) = 0.85, rho(xy) = 0.80........... 108

3.11 n=25, rho(xz) = 0.94, rho(yz) = 0.85, rho(xy) = 0.80........... 109

3.12 n=50, rho(xz) = 0.94, rho(yz) = 0.85, rho(xy) = 0.80........... 110

3.13 n=10, rho(xz) = 0.95, rho(yz) = 0.95, rho(xy) = 0.90........... 111

3.14 n=25, rho(xz) = 0.95, rho(yz) = 0.95, rho(xy) = 0.90........... 112

3.15 n=50, rho(xz) = 0.95, rho(yz) = 0.95, rho(xy) = 0.90........... 113

3.16 n=10, rho(xz) = 0.97, rho(yz) = 0.97, rho(xy) = 0.95........... 114

3.17 n=25, rho(xz) = 0.97, rho(yz) = 0.97, rho(xy) = 0.95........... 115

3.18 n=50, rho(xz) = 0.97, rho(yz) = 0.97, rho(xy) = 0.95........... 116

3.19 n=10, rho(xz) = 0.00, rho(yz) = 0.10, rho(xy) = 0.95........... 117

3.20 n=25, rho(xz) = 0.00, rho(yz) = 0.10, rho(xy) = 0.95........... 118

3.21 n=50, rho(xz) = 0.00, rho(yz) = 0.10, rho(xy) = 0.95........... 119

3.22 n=10, rho(xz) = 0.92, rho(yz) = 0.65, rho(xy) = 0.88........... 120

3.23 n=25, rho(xz) = 0.92, rho(yz) = 0.65, rho(xy) = 0.88........... 121

3.24 n=50, rho(xz) = 0.92, rho(yz) = 0.65, rho(xy) = 0.88........... 122

3.25 n=10, rho(xz) = 0.93, rho(yz) = 0.75, rho(xy) = 0.92........... 123

3.26 n=25, rho(xz) = 0.93, rho(yz) = 0.75, rho(xy) = 0.92............ 124
3.27 n=50, rho(xz) = 0.93, rho(yz) = 0.75, rho(xy) = 0.92............ 125
3.28 n=10, rho(xz) = 0.94, rho(yz) = 0.85, rho(xy) = 0.96............ 126
3.29 n=25, rho(xz) = 0.94, rho(yz) = 0.85, rho(xy) = 0.96............ 127
3.30 n=50, rho(xz) = 0.94, rho(yz) = 0.85, rho(xy) = 0.96............ 128
3.31 n=10, rho(xz) = 0.95, rho(yz) = 0.95, rho(xy) = 0.98............ 129
3.32 n=25, rho(xz) = 0.95, rho(yz) = 0.95, rho(xy) = 0.98............ 130
3.33 n=50, rho(xz) = 0.95, rho(yz) = 0.95, rho(xy) = 0.98............ 131
3.34 n=10, rho(xz) = 0.97, rho(yz) = 0.97, rho(xy) = 0.99............ 132
3.35 n=25, rho(xz) = 0.97, rho(yz) = 0.97, rho(xy) = 0.99............ 133
3.36 n=50, rho(xz) = 0.97, rho(yz) = 0.97, rho(xy) = 0.99............ 134
3.37 n=25, rho(xz) = 0.00, rho(yz) = 0.10, rho(xy) = 0.50............ 135
3.38 n=25, rho(xz) = 0.92, rho(yz) = 0.65, rho(xy) = 0.75............ 136
3.39 n=25, rho(xz) = 0.93, rho(yz) = 0.75, rho(xy) = 0.80............ 137

Scatterplot For Correlation Estimates

3.40 n=25, rho(xz) = 0.00, rho(yz) = 0.10, rho(xy) = 0.00............ 138
3.41 n=25, rho(xz) = 0.93, rho(yz) = 0.75, rho(xy) = 0.92............ 139
3.42 n=25, rho(xz) = 0.00, rho(yz) = 0.10, rho(xy) = 0.00............ 140
3.43 n=25, rho(xz) = 0.00, rho(yz) = 0.10, rho(xy) = 0.00............ 141
3.44 n=25, rho(xz) = 0.93, rho(yz) = 0.75, rho(xy) = 0.92............ 142
3.45 n=25, rho(xz) = 0.93, rho(yz) = 0.75, rho(xy) = 0.92............ 143

CHAPTER 1. INTRODUCTION AND SUMMARY

One of the most important tools for analyzing economic policies is the micro-analytic model. It is used frequently in reaching public policy decisions. Virtually every federal agency employs micro-analytic models for the evaluation of policy proposals.

Direct use of sample observations rather than aggregated data is characteristic of the micro-analytic approach. For this reason, the type of micro-data that is used as input to the model has a significant bearing on the validity of the results of the model. Furthermore, when all the input-data come from a single sample, the quality of the model depends on, among others, sampling and data-recording procedures.

However, if the data from a single source is insufficient or partly aggregated then, typically, multiple sources of data are used to provide the necessary input to the model. In such situations, researchers use a methodology in which multiple sources of data are merged to form a composite data-file. Assessing the validity and quality of the results from the model for the composite data-file becomes more difficult than from the models for a single source data-file. Effective use of data from multiple sources in order to produce comprehensive files which lead to meaningful statistical usage is the fundamental issue in the file-merging methodology.

Save for the record-linkage problem (see Section 1.1), which forms only a part of the general problem of merging files of data, the statistical properties of the methodology are largely unexplored.

The report of the Federal Subcommittee on Matching Techniques [see Radner et.al (1980)] presents some theoretical foundation and empirical justification for the file-merging methodology. Further, recent effort in this direction is presented in Ramalingam (1985), Rodgers (1984), Rubin (1986) and Woodbury (1983). This monograph reviews the relevant literature and then presents new statistical properties of some known procedures for merging data-files.

1.1. A Dichotomy of Matching Problems.

In general, the following categories of file-matching problems have been distinguished [DeGroot (1987) and Radner et.al.(1980)] : (i) problems of *record-linkage* or *exact-matching* in which it is desired to identify pairs of records in two files that pertain to the same individual, and (ii) problems of *statistical matching* in which the goal is to identify pairs of records, one from each of two files, that correspond to similar individuals. After describing these problems briefly, we delineate the class of matching problems that shall be pursued in this monograph by giving a broad interpretation to the phrase *statistical matching* .

Record-linkage is contemplated when identifiers such as social security number, name, address, etc. are available to perform matching of records in the two files. In such situations, all we need is an efficient software to sort the individuals by their identifiers. With the help of such software, we can, within reasonable error, link a given individual in one file with an individual in the other file such that these two units correspond to the same values for the identifiers. The resulting merged file contains micro-data which are more comprehensive than the two separate files. While most records in the merged file pertain to matching of the separate records of an individual, the number of erroneous matches that put records of different people together in the enlarged file depends partly on (a) the particular software used in the process of merging, and (b) the accuracy of information on identifiers. For these reasons, two records that pertain to an individual may not be paired with each other. To minimize these mis-matches and non-matches, various models and techniques have been proposed. Notable among these are the models proposed by Fellegi and Sunter (1969) and Dubois (1969). Among the well-known exact-match studies, mention must be made of the Framingham heart study reported in Dawber et al

(1963) and the study of Japanese A-bomb survivors discussed in Beebe (1979). The rich history of record-linkage research and commentaries on current activities in exact-matching of files of data, including an up-to-date bibliography on exact-match studies, have been well documented in Kilss et al (1985). It is clear that if accurate identifiers are available for the units in the two files, then no statistical issues are involved in the record-linkage methodology. However, such exact-matching is often not possible because of various reasons discussed below.

Over the past several years, there have been significant changes in the laws and regulations pertinent to exact matching of records for statistical and research purposes. New laws, especially the Privacy Act of 1974 and the Tax Reform Act of 1976, have imposed additional restrictions on the matching of records belonging to more than one federal agency and on the matching of files of federal agencies with those of other organizations. As a result, some agencies tend to limit access to their records to an even greater extent than seems necessary by statutory requirements.

Analyses of micro-economic models often involve data that are available only from multiple sources. For example, suppose that one is interested in the relationships among two sets of variables, one set consisting of information about health care expenses incurred by individuals and the other set consisting of information about receipt of various types of welfare benefits. Suppose further that no existing data file contains all of the needed variables, but that two sample surveys of a target population together contain all these variables. If executing a new survey to obtain all the variables from a single sample is not feasible, then one might match the two samples and use the merged file for statistical analyses of variables which are not present in the same sample. The two samples may have information on the *same* individuals or else the two samples are stochastically independent, i.e., in practice they contain very few or no individuals in common. In the former case, reliable identifiers may not be available, whereas in the latter case identifiers are clearly useless for the creation of merged file.

In these situations, the record linkage methodology is inadequate for the purpose of merging the two files of data. However, assumed probabilistic models for the mechanism to generate the data in the two files provide the framework in which statistical techniques can often be used to combine these files. Such procedures will be called *statistical matching* strategies and the resulting merged file will be called a *synthetic file*. Some models and specific strategies for statistical matching are discussed in Kadane (1978), Rubin (1986), and DeGroot (1987). In this monograph, we shall not discuss record-linkage problems based on identifiers any further.

It is customary to distinguish between matching strategies required when *same* (exact matching) or *similar* (statistical matching) individuals are in the two files, see DeGroot (1987). However, in the literature on matching files, there is no consensus on definitions of exact match and statistical match. For example, Woodbury (1983) differs from the usual practice in the sense that procedures for merging files involving the same individuals have been described as "statistical record matching for files." It may be noted that given the records of the same people, matching done by statistical methodologies, in the absence of identifiers, do not lead to exact/perfect matches. Therefore, in this monograph we shall call any model-based statistical technique for merging two files as a statistical matching strategy, regardless of whether the files consist of the same individuals or similar individuals.

1.2 A Paradigm

A micro-economic model that has been used at the Office of Tax Analysis (OTA), US Department of Treasury, is the Federal Personal Income Tax Model [see Barr and Turner (1978)]. This model is used to assess proposed tax law changes in terms of their effects on the distribution of after-tax income, the efficiency with which the changes will operate in achieving their objectives, etc. The inputs for this model are two sources of micro-data, namely the Statistics of Income File (SOI) and the Current Population Survey (CPS). The SOI file is generated annually by the Internal Revenue Service (IRS) and it consists of personal tax return data. The CPS file is produced monthly by the Bureau of the Census in which the reporting units are households. As explained in Section 1.1, such pooling of data from more than one Federal Agency has been severely restricted in recent years by confidentiality issues arising because of the privacy laws. For this reason, complete information on individuals, especially identifiers such as social security numbers, is typically not released by the IRS and the Census Bureau. The resulting micro-data files are a compromise between complete Census files and fully aggregated data-sets. In other words, sufficient detail remains to support micro-analysis of the population, while partial aggregation protects individual privacy and greatly diminishes computational burden.

A typical problem in tax-policy evaluation occurs when no single available data file such as SOI or CPS contains all the information needed for an analysis. For example, consider the variables in $\mathbf{W} = (X, Y, Z_1, Z_2)$, where

X = Allowable itemizations and capital gains

Y = Old Age Survivors Disability Insurance (OASDI)

Z_1 = Social security number

Z_2 = Marital status

Suppose that we are interested in estimating the correlation ρ_{xy} between X and Y or, more generally, the expectation (assumed finite) of a known function g, say, of **W**; that is the integral

$$\gamma = \int g(w)\, dF(w), \tag{1.2.1}$$

where F(w) is the joint distribution function of the variables in **W**. Now, the SOI micro data file cannot be used in its original form since it does not include the OASDI benefits (Y). Census files (CPS) with OASDI benefits do not allow a complete analysis of the effect of including this benefit, since it does not contain information on allowable itemizations and capital gains (X). Thus, instead of observing X, Y, Z_1, Z_2 jointly on the same units, we can only get the following pair of files:

File 1 (SOI): containing X, Z_1, Z_2

and

File 2 (CPS): containing Y, Z_1, Z_2 .

Estimating γ based on the fragmentary data provided by File 1 and File 2 is an important practical problem that has not yet been solved satisfactorily. In an attempt to cope with situations such as the OTA model, federal agencies have long been using procedures for matching or merging the two incomplete files so that one can carry out the usual inference for g, hoping that the merged file is a reasonable substitute for the unobserved data on (X, Y, Z_1, Z_2).

1.3. A General Set-up for Statistical Matching

Consider a universe $\mathcal{U}$ of individuals. Let **X, Y, Z** denote three groups of random variables and let us assume that we cannot observe the vector **W** = (**X, Y, Z**) for any unit in $\mathcal{U}$. Assume that none of these variables in an identifier, thus ruling out exact matching based on identifiers. However, suppose that the following data are available:

File 1 (Base): n_1 individuals, each with information on some function, say, $\mathbf{W}_1^*$ of **W**, and

File 2 (Supplementary): n_2 ($\geq n_1$) individuals, each with information on another function, say, $\mathbf{W}_2^*$ of **W**.

Statistical Matching of File 1 and File 2
with information on

↓	↓	↓
$\mathbf{W}_1^* = \mathbf{X}$	$\mathbf{W}_1^* = (\mathbf{X}, \mathbf{Z} + \underset{\sim}{\varepsilon}_1)$	$\mathbf{W}_1^* = (\mathbf{X}, \mathbf{Z})$
and	and	and
$\mathbf{W}_2^* = \mathbf{Y}$	$\mathbf{W}_2^* = (\mathbf{Y}, \mathbf{Z} + \underset{\sim}{\varepsilon}_2)$	$\mathbf{W}_2^* = (\mathbf{Y}, \mathbf{Z})$
for same individuals.	for same individuals, where $\underset{\sim}{\varepsilon}_1$ and $\underset{\sim}{\varepsilon}_2$ are possible independent measurement errors.	for similar individuals.

Figure 1.1.

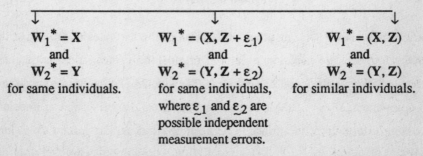

Various matching problems arise depending on what type of data are in $\mathbf{W}_1^*$ and $\mathbf{W}_2^*$. We distinguish between only three different situations as discussed in Figure 1.1.

1.4. The Statistical Matching Methodology

We shall now mention some important steps involved in actually creating a statistical match between two given files (see Radner et al, 1980). First, the basic assumption in the general framework outlined in Section 1.3, that the incomplete files of data come from a common population of units, may sometimes be violated. In that case, a "universe adjustment" needs to be carried out to ensure that there is a common universe $\mathcal{U}$ from which individuals of the files are sampled. Second, a "units adjustment" might be needed if the units of observation in the two files differ (e.g., persons and tax units). Third, matching or overlapping variables, **Z**, are identified and it is assumed that File 1 with n_1 records carries information on (**X**, **Z**), whereas File 2 with n_2 records consists of data on (**Y**, **Z**).

Finally, the "merging" step is carried out differently in the various scenarios outlined in Figure 1.1. Taking first the situation of matching files with the *same* individuals into consideration, if at least one overlapping variable is continuous and is free of measurement errors, then it can be effectively used as an identifier because probability of a tie for such a variable is zero. Thus, in this special case, no statistical matching is needed to put together the two files. The record linkage models or exact matching (as in Radner et al, 1980, Chapter II) can be adopted. Also, if every overlapping variable is a discrete or categorical variable, then they serve as 'cohort' variables (see Woodbury (1983)) and one can create 'packets' of subfiles with no variation in the cohort variable. For example, if sex is the cohort variable in both the files, then a pair of subfiles, one each from File 1 and File 2 can be created, in which the units are males and another pair of subfiles in which the units are females. Once packets are created with regard to all the **Z** variables, then matching needs to be done inside each packet between subfiles with **X** information only and subfiles with **Y** information only. Thus, in this particular case of only categorical variables in

Z, one can reduce the matching problem to one involving no overlapping variable (see Figure 1.1) Matching in such contexts have been addressed by many investigators [see DeGroot (1980) and references cited therein] who have proposed both frequentist and Bayesian matching strategies assuming a probability model for (**X**, **Y**).

When continuous overlapping variables are measured with errors, and some overlapping variables are discrete, then packets of records are created in order to diffuse the dissimilarities between units that are being matched. The selection of "matching records" at the packet level is typically based on some measure of distance between a given record in File 1 and each potential match in the (Supplementary) File 2. A potential match with the smallest distance (see Section 1.5) is chosen as the match that will provide the missing **Y** value to a File 1 record. Thus we obtain the synthetic file containing the data

$$(\mathbf{X}_i, \mathbf{Y}_{j(i)}, \mathbf{Z}_i) \text{ , } i = 1,..., n_1,$$

where j(i) is the index of the potential match record from File 2 that is imputed on to ($\mathbf{X}_i$, $\mathbf{Z}_i$) in

File 1.

1.5 Constrained and Unconstrained Matching

In the last section, we mentioned 'distance' based matching of records from two files. Depending on the number of records from File 2 which are matched with a record from File 1, two different types of matching strategies are distinguished (See Radner et.al. 1980, p. 18). To fix our ideas, suppose that the age of an individual, Z_1, say, is a variable overlapping the two files. Then one may define a distance measure d, say, between the ith individual in File 1 and the jth individual in File 2 by the equation

$$d_{ij} = | Z_{1i} - Z_{2j} | .$$

For fixed $i = 1, 2,..., n_1$, one will then match one possible j^* in File 2 with the ith record in File 1 if j^* minimizes d_{ij} over j. That is j^* depends possibly on i and satisfies the equation

$$d_{ij^*} = \min_{1 \le j \le n_2} d_{ij}$$

If the choice of j* involves no further restrictions, so that the same j* can be matched with multiple records from File 1, then the statistical matching strategy is called *unconstrained matching*.

However, usually there are additional restrictions, subject to which one must choose the optimal match j* from File 2. For example, even though unconstrained matchings allow the closest possible match for each record in File 1, the joint distributional properties of $(\mathbf{Y}, \mathbf{Z})$ variables in File 2, such as variance-covariance matrices, may not be preserved in the merged file obtained by *unconstrained matchings*.

In order to avoid such disadvantages in the merged file, *constrained matching* has been proposed [e.g., Barr and Turner (1980), Hollenbeck & Doyle (1979)]. A mathematical formulation of this type of matching can be given as follows. Assume first that both files contain only n records, that is the common value of n_1 and n_2 is n. For $1 \le i, j \le n$, let

$$a_{ij} = \begin{cases} 1 & \text{if records i in File 1 and j in File 2 are matched} \\ 0 & \text{otherwise} \end{cases} \qquad (1.5.1)$$

Then, the following additional conditions will ensure that the aforementioned moment properties are preserved by not letting more than one record in File 1 to be matched with the same record in File 2:

$$\sum_{i=1}^{n} a_{ij} = 1, \quad \text{for } j = 1, 2,..., n \qquad (1.5.2)$$

$$\sum_{j=1}^{n} a_{ij} = 1, \quad \text{for } i = 1, 2,..., n \qquad (1.5.3)$$

Now let d_{ij} denote a measure of inter-record dissimilarity given by the extent to which the attributes in any one record differ from the attributes in another record. Then the optimal constrained match minimizes the objective function

$$\sum_{i=1}^{n} \sum_{j=1}^{n} d_{ij} \, a_{ij}, \tag{1.5.4}$$

subject to the restrictions in (1.5.2) and (1.5.3). Clearly, this extremal problem is the standard linear assignment problem in mathematical programming.

A matching situation more typical of problems relating to policy analyses is a constrained merge of two files with variable weights for records in both files and an unequal number of records in the files. Let α_i be the weight of the ith record in File 1, and let β_j be the weight of the jth record in File 2. If n_1, n_2 are respectively, the number of records in File 1 and File 2, then we minimize the objective function in (1.5.4) subject to the following constraints.

$$\sum_{j=1}^{n_2} a_{ij} = \alpha_i, \; i = 1, 2, \ldots, n_1 \tag{1.5.5}$$

$$\sum_{i=1}^{n_1} a_{ij} = \beta_j, \; j = 1, 2, \ldots, n_2 \tag{1.5.6}$$

$$\sum_{i=1}^{n_1} \alpha_i = \sum_{j=1}^{n_2} \beta_j \tag{1.5.7}$$

and

$$a_{ij} \geq 0, \; \forall \text{ i and j}. \tag{1.5.8}$$

It is clear that an optimal constrained matching strategy when the two files have unequal number of individuals is the solution of a standard transportation problem in which the roles of the "warehouses" and "markets" are respectively played by the records in File 1 and File 2 and the "cost of transportation" is the inter-record distance "d_{ij}." Existing algorithms to solve a linear assignment or transportation problem can be used to complete the final "merge" step, giving us the synthetic sample

$$\mathbf{W}_i^* = (\mathbf{X}_i, \mathbf{Y}_j^*, \mathbf{Z}_i)\ ,\ \text{with weights } a_{ij}^*,\ 1 \le i \le n_1 \tag{1.5.9}$$

where $\mathbf{Y}_j^*$ denotes the value of $\mathbf{Y}$ assigned to the ith record of File 1 with weight a_{ij}^*. The sample in (1.5.9) may now be used to estimate a parameter like γ in (1.2.1).

1.6 Reliability of Synthetic Files

The precision of synthetic-file based estimators of a given parameter relevant to the population of $\mathbf{W} = (\mathbf{X}, \mathbf{Y}, \mathbf{Z})$ is affected by various types of errors that occur while matching two files. To discuss these matching errors, let us first restrict our attention to the cases where the *same* individuals are in the two files.

Due to uncertainties, it is inevitable that some matching errors occur, even with the most sophisticated merging procedure. These errors fall into two major categories:

(i) Erroneous match (false match) or linking of records that correspond to different individuals.

(ii) Erroneous non-match (false non-match) or failure to link the records that do correspond to the same individual.

The reliability of the results of a statistical matching strategy can be defined as one of the following coefficients:

(a) the proportion (or the count) of the correct matches, that is, matches of records on the same individuals.

(b) the proportion (or the count) of erroneous decisions, that is, false matches and erroneous non-matches.

These reliability coefficients are random variables because a statistical matching strategy is dependent on the data in the two files. The sampling distribution of the reliability coefficients, either exact or asymptotic (as the sizes of the files grow), are very useful in judging the quality of a given matching procedure. If the exact sampling distribution of the reliability coefficient is intractable, then one can study suitable summary measures such as the expected number of erroneous (or correct) matches in order to evaluate a matching strategy.

The coefficients (a) and (b) mentioned above are typical measures of reliability in the context of record linkage problems [see Radner et.al (1980), p. 13] and in the context of statistical matching for same individuals [see DeGroot, Feder and Goel (1971)].

Now, we will discuss the reliability of a synthetic file when the two files contain similar individuals. First note that the matching error defined in the case of same individuals, are not applicable in file-matching for similar individuals because the linkage of records that pertain to the same unit is usually not possible. Thus the number of correct matches is irrelevant to assessing the quality of the synthetic file when similar individuals are involved in the two files. Not many definitions of error and reliability, which are tractable from a theoretical point of view, are available. Kelley (1983) examines the error structure of synthetic files created by merging files on similar individuals. His approach is motivated by the fact that almost all matching procedures cited in the literature (see Section 3.2) make the implicit assumption that X and Y are conditionally independent given the value of Z, where (X, Z) and (Y, Z) information come separately from different files. Restricting his attention to errors induced by the violation of this assumption, he proposes the ratio

$$f(x \mid y, z) \,/\, f(x \mid z),$$

as a measure of the error in the matched sample due to such violations where $f(x \mid y,z)$ is the conditional density of x given y and z and $f(x \mid z)$ is the conditional density of x given z. Assuming that the unobserved triple (X,Y, Z) is a trivariate normal vector, he derives upper bounds for this error measure. Since these bounds are expressed in terms of parameters pertaining to the distributions of the observed vectors (X, Z) and (Y, Z), severity of the errors in the matched sample due to violations of the conditional independence assumption can be understood to some extent in this specific tri-variate normal model.

1.7 Summary

The problem of merging micro-data files, various steps involved in the existing methodology to solve this problem, and related issues have been described in this chapter. Specifically, a general framework to categorize knownmodels for matching files has been proposed in this chapter. This framework includes situations in which files of data on either same or similar individuals need to be merged. This monograph is concerned with both theoretical investigations and empirical evaluations of the quality of synthetic files.

In Chapter 2, situations pertinent to merging files on the same individuals are treated. A review of known results in this direction is given. New optimality properties of a maximum likelihood matching strategy are established. Some small-sample and large-sample properties of the number of correct matches, which shed some light on the reliability of the synthetic file arising from using the maximum likelihood strategy, are derived.

Chapter 3 deals with the situation of merging data on similar individuals. The bulk of the discussion in this chapter is confined to matching two files of data that are sampled from a trivariate normal population. Thus, if (X, Y, Z) is a three-dimensional normal random vector, File 1 has data on (X, Z), while File 2 has data on (Y, Z). Two strategies proposed by Kadane (1978) and one strategy due to Sims (1978) are used to create synthetic files out of simulated data on (X, Z) and (Y, Z). These synthetic files are then evaluated by comparing the estimates of the correlation between X and Y based on them with the estimates based on unbroken data (X, Y, Z).

CHAPTER 2. MERGING FILES OF DATA ON SAME INDIVIDUALS

A useful classification of situations involving statistical matching of data files was discussed in Section 1.3. It may be recalled that in the context of the two files comprising the same individuals, this classification scheme included two scenarios. One where no overlapping variables are present, and the other where overlapping variables are available. We begin with the former situation, relegating the latter to Section 2.10.

2.1 A General Model

Let $\mathbf{W}' = (\mathbf{T}', \mathbf{U}')$ be a multi-dimensional random vector with CDF $H(\mathbf{t},\mathbf{u})$ and PDF $h(\mathbf{t},\mathbf{u})$. Let $\mathbf{W}_i' = (\mathbf{T}_i', \mathbf{U}_i')$, $i = 1,2,\ldots, n$ be a random sample of size n from this population. We shall assume that these sample values got broken up into the component vectors **T**'s and **U**'s before the data could be recorded. Thus we do not know which **T** and **U** values were paired in the original sample and the two files consist of the following data:

File 1 - $\mathbf{X}_1,\mathbf{X}_2,\ldots,\mathbf{X}_n$, which is an unknown permutation of $\mathbf{T}_1,\mathbf{T}_2,\ldots,\mathbf{T}_n$, and

File 2 - $\mathbf{Y}_1, \mathbf{Y}_2,\ldots, \mathbf{Y}_n$, which is an unknown permutation of $\mathbf{U}_1,\ldots, \mathbf{U}_n$. DeGroot, Feder and Goel (1971) call this a *Broken Random Sample* model for two files.

Two types of statistical decision and inference problems arise from observing a broken random sample. The first type of problem involves trying to pair the **x**'s with the **y**'s in the broken data in order to reproduce the pairs in the original unbroken sample. The second type of problem involves making inferences about the values of parameters in the joint distribution $H(\mathbf{t}, \mathbf{u})$ of **T** and **U**.

This chapter will be organized into a review of the literature on matching problems in Sections 2.3 to 2.5, followed by a discussion of statistical properties of some matching strategies in Sections 2.6 to 2.9.

2.2 Notations

In this section, we introduce some of the notations that will be used in the present chapter.

(i) The random vector **W** is assumed to have an absolutely continuous joint CDF H(**t**, **u**) and joint density h(**t**, **u**). In particular, $(T, U)'$ will denote a two-dimensional random vector, with h(t, u) and H(t, u) respectively as the density and CDF of $(T, U)'$, and $F(\cdot)$, $G(\cdot)$ the respective marginal distribution functions of T and U. The symbol $g_\xi(\cdot)$ will be the generic notation for the density function of the random vector ξ. Without the suffix, $g(\cdot)$ will denote an arbitrary real-valued function.

(ii) Let $\mathbf{W}_i$, $1 = 1, 2, \ldots, n$ be a random sample from the population of **W**. Let $F_n(x) = \frac{1}{n} \sum_{i=1}^{n} I_{(T_i \leq x)}$ denote the empirical DF based on the variables $T_1, \ldots, T_n$ [where I_A means the indicator of the event A]. Similarly, $G_n(x)$ will be the empirical DF based on $U_1, \ldots, U_n$. For $i = 1, 2, \ldots, n$, let R_i denote the rank of T_i among the variables $T_1, \ldots, T_n$, and S_i denote the rank of U_i among the variables $U_1, \ldots, U_n$.

(iii) Let $\varphi = (\varphi(1), \ldots, \varphi(n))$ be a permutation of the integers $1, 2, \ldots, n$. Φ will stand for the set all such permutations. Also, let $\varphi^* = (1, 2, \ldots, n)$.

(iv) Let $\varepsilon \geq 0$. For all $i = 1, 2, \ldots, n$ define events $A_{ni}(\varphi, \varepsilon)$, $A_{ni}(\varepsilon)$ and A_{ni} as follows:

$$A_{ni}(\varphi, \varepsilon) = [\, | U_{(\varphi(R_i))} - U_i | \leq \varepsilon] \tag{2.2.1}$$

Let $A_{ni}(\varepsilon) = A_{ni}(\varphi^*, \varepsilon)$, $i = 1, 2, \ldots, n$, (2.2.2)

$$A_{ni} = A_{ni}(\varphi^*, 0) \equiv \{R_i = S_i\}. \tag{2.2.3}$$

Let $\upsilon_{ni}(\varphi, \varepsilon) = I_{A_{ni}(\varphi, \varepsilon)}$, $i = 1, 2, \ldots, n$. (2.2.4)

$$\upsilon_{ni}(\varepsilon) = I_{A_{ni}}(\varphi^*,\varepsilon),\ i = 1, 2,\ldots, n. \tag{2.2.5}$$

$$\upsilon_{ni} = I_{A_{ni}},\ i = 1, 2,\ldots, n. \tag{2.2.6}$$

(v). For any fixed integer $d \le n$, define

$$\mathbf{B}_n = (B_{n1},\ldots, B_{nd})', \text{ where} \tag{2.2.7}$$

$$B_{nj} = R_j - S_j,\ j = 1, 2,\ldots, n.$$

Note that if $\forall\ 1 \le j \le d$ and $1 \le k \le n$,

$$\xi_{jk} = I_{(T_j - T_k \ge 0)} - I_{(U_j - U_k \ge 0)}, \tag{2.2.8}$$

and $\underset{\sim}{\xi}_k\ (\xi_{1k}, \ldots, \xi_{dk})'$, then we get the representation

$$B_{nj} = \sum_{k=1}^{n} \xi_{jk},\ j = 1, 2, \ldots, d. \tag{2.2.9}$$

and

$$\mathbf{B}_n = \sum_{k=1}^{n} \underset{\sim}{\xi}_k. \tag{2.2.10}$$

(vi). Let Λ_d be the sigma-field $\sigma(\mathbf{W}_1,\ldots, \mathbf{W}_d)$ generated by the vectors $\mathbf{W}_i$, $i = 1, 2,\ldots, d$. Let $\psi_{\underset{\sim}{\eta}}(\underset{\sim}{\theta})$ be the generic notation for the characteristic function of a random vector $\underset{\sim}{\eta}$, $\underset{\sim}{\theta}$ being a vector of dummy variables whose dimension is the same as that of $\underset{\sim}{\eta}$.

(vii). Let $\xi_{jk}(\mathbf{w}_1,\ldots, \mathbf{w}_d)$, $\underset{\sim}{\xi}_k\ (\mathbf{w}_1,\ldots, \mathbf{w}_d)$ and $\mathbf{B}_n\ (\mathbf{w}_1,\ldots, \mathbf{w}_d)$ be respectively ξ_{jk}, $\underset{\sim}{\xi}_k$ and $\mathbf{B}_n$ given that $\mathbf{W}_i = \mathbf{w}_i$, $i = 1, 2,\ldots,d$.

(viii). Let $\psi_d = \psi_d\ (\mathbf{w}_1,\ldots, \mathbf{w}_d)$ be the negative logarithm of the modulus of the characteristic function of $\underset{\sim}{\xi}_{d+1}\ (\mathbf{w}_1,\ldots, \mathbf{w}_d)$.

2.3 Model-based Matching Strategies

Pairing the observations in the two data-files, described in Section 2.1, should be distinguished from the problem of matching two equivalent decks of n distinct cards, which is discussed in elementary textbooks such as Feller (1968). One version of card-matching is as follows. Consider a "target pack" of n cards laid out in a row and a "matching pack" of the same number of cards laid out *randomly* one by one beside the target pack. In this random arrangement of cards, n pairs of cards are formed. A match or coincidence is said to have occurred in a pair if the two cards in the pair are identical. Because the two decks are merged purely by chance and without using any type of observations or other information about the cards, one may describe such problems as *no-data matching problems*. An excellent survey of various versions of card-matching schemes is found in Barton (1958) and David and Barton (1962, Chapter 12)

Suppose that N denotes the number of pairs in the card-matching problem which have like cards or matches. The derivation of the probability distribution of N dates back to Montmort (1708). The following is a summary of some of the well-known properties of N (Feller 1968):

Proposition 2.3.1. *If* $P_{[m]}$ *is the probability of having exactly m matches, then*

(i) $$P_{[m]} = \frac{1}{m!}[1 - 1 + \frac{1}{2!} - \frac{1}{3!} + \dots \pm \frac{1}{(n-m)!}]\ , \ m = 0, 2, \dots, n-1$$

and

$$P_{[n]} = \frac{1}{n!}$$

(ii) *Noting that* $\frac{e^{-1}}{m!}$ *is the probability that a Poisson random variable with mean 1 takes the value m, we have the following approximation for large n:*

$$P_{[m]} \approx \frac{e^{-1}}{m!}$$

(iii) *For* $d = 1, 2, \dots, n$, *the dth factorial moment of N,* $E(N^{(d)})$, *is 1.*

In Section 2.9, it will be shown that the optimal matching strategy based on independent variables T and U is only as good as the no-data matching. However, as one might expect, for certain broken random sample models, it pays to match two files of data using optimal strategies based on such data. Several authors starting with DeGroot, Feder and Goel (1971) have proposed and studied matching strategies based on models for broken data.

2.4 Repairing a Broken Random Sample

2.4.1 The Basic Matching Problem

Let us consider matching the broken random sample $x_1, x_2, \ldots, x_n, y_1, \ldots, y_n$ by pairing x_i with $y_{\varphi(i)}$, for $i = 1, 2, \ldots, n$ where $\varphi = (\varphi(1), \ldots, \varphi(n))$ is a permutation of 1, 2,..., n. As we seek a $\varphi \in \Phi$ that will provide reasonably good pairings of the x's with the y's, we need to clarify the fundamental role of φ in the statistical model described in Section 2.1. If we treat φ as an unknown parameter of the model, then the likelihood of the data will include φ. For instance, if T and U are jointly bivariate normal with means μ_1, μ_2, variances σ_1^2, σ_2^2 and correlation coefficient ρ, then the log-likelihood function of $\varphi, \rho, \mu_1, \mu_2, \sigma_1, \sigma_2$, given the broken random sample, is

$$\mathcal{L}(\varphi, \rho, \mu_1, \mu_2, \sigma_1, \sigma_2 \mid x_1, \ldots, x_n, y_1, \ldots, y_n)$$

$$= -\frac{n}{2}\log(1-\rho^2) - n\log(\sigma_1, \sigma_2)$$

$$- \frac{1}{2(1-\rho^2)}\Big[\frac{1}{\sigma_1^2}\sum_{i=1}^{n}(x_i-\mu_1)^2 + \frac{1}{\sigma_2^2}\sum_{i=1}^{n}(y_i-\mu_2)^2$$

$$- \frac{2\rho}{\sigma_1\sigma_2}\sum_{i=1}^{n}(x_i-\mu_1)(y_{\varphi(i)}-\mu_2)\Big]. \qquad (2.4.1)$$

A constant term not involving the parameters has been omitted in (2.4.1). In the following subsection, we shall seek φ's that maximize this likelihood. However, some statisticians would regard φ as some sort of missing data and not as a parameter of the underlying model. The problem of pairing the two files will not arise in such situations. However, one may still want to carry out statistical inference for other parameters of the model based on the broken random sample. Such issues are not pursued here and one may refer to DeGroot and Goel (1980) for an approach to estimating the correlation coefficient ρ while treating φ as missing data in the bivariate normal model.

2.4.2 The Maximum Likelihood Solution to the Matching Problem

We start with a bivariate model used in DeGroot et al. (1971) which assumes that the bivariate parent probability density function of **W** is

$$h(t,u) = \alpha(t)\, \beta(u) \exp[\gamma(t)\, \delta(u)] \qquad (2.4.2)$$

where α, β, γ, δ are *known* but otherwise arbitrary real valued functions of the indicated variables. Suppose now that $x_1, \ldots, x_n$ and $y_1, \ldots, y_n$ are the observations in a broken random sample from a completely specified density of the form (2.4.2). If x_i was paired with $y_{\varphi(i)}$ for $i = 1, 2, \ldots, n$, in the original unbroken sample, then the joint density of the broken sample would be

$$\prod_{i=1}^{n} h[x_i, y_{\varphi(i)}] = [\prod_{i=1}^{n} \alpha(x_i)][\prod_{i=1}^{n} \beta(y_i)] \exp[\sum_{i=1}^{n} \gamma(x_i)\, \delta(y_{\varphi(i)})] \,. \qquad (2.4.3)$$

Thus the maximum likelihood estimate of the unknown φ is the permutation for which $\sum_{i=1}^{n} \gamma(x_i)\, \delta(y_{\varphi(i)})$ is maximum. Without loss of generality, we shall assume that the x_i's and y_j's have been reindexed so that $\gamma(x_1) \le \ldots \le \gamma(x_n)$ and $\delta(y_1) \le \ldots \le \delta(y_n)$. Since **W** is assumed to have an absolutely continuous distribution, there are no ties among $\gamma(x_i)$'s or $\delta(y_j)$'s. DeGroot et al (1971) show that the maximum likelihood solution is to pair x_i with y_i, for $i = 1, \ldots, n$. In other words, the maximum likelihood pairing (M.L.P.) is $\varphi^* = (1, \ldots, n)$.

In particular, if the density in (2.4.2) is that of a bivariate normal random vector with correlation ρ, then the M.L.P. can be described knowing only the sign of ρ. If $\rho > 0$, the M.L.P. is to order the observed values so that $x_1 < \ldots < x_n$ and $y_1 < \ldots < y_n$ and then to pair x_i with y_i, for $i = 1, 2, \ldots, n$. If $\rho < 0$, the solution is to pair x_i and $y_{(n+1-i)}$, for $i = 1, 2, \ldots, n$. If $\rho = 0$, all pairings, or permutations, are equally likely.

Chew (1973) derived the maximum likelihood solution to the (bivariate) matching problem for a larger class of densities h(t,u) with a monotone likelihood ratio. That is, for any values t_1, t_2, u_1 and u_2 such that $t_1 < t_2$ and $u_1 < u_2$,

$$h(t_1,u_1)\ h(t_2,u_2) \geq h(t_1,u_2)\ h(t_2,u_1) \tag{2.4.4}$$

Let the values $x_1,\ldots, x_n$ and $y_1,\ldots, y_n$ in a broken random sample be from a density h(t,u) satisfying (2.4.4). Without loss of generality, relabel the x's and y's so that $x_1 < \ldots < x_n$ and $y_1 < \ldots < y_n$. Then permutation $\varphi^* = (1,\ldots, n)$ is again the M.L.P.

2.4.3 Some Bayesian Matching Strategies

DeGroot et al. (1971) studied the matching problem from a Bayesian point of view as well. They proposed three optimality criteria, subject to which one may choose the matching strategy φ. Before stating these criteria, some notation and definitions are needed.

Let $x_1 < \ldots < x_n$ and $y_1 < \ldots < y_n$ be the values of a broken random sample from a given parent distribution with density h(t,u). If x_i is paired with $y_{\varphi(i)}$, $i = 1, 2, \ldots, n$, then the likelihood function of the unknown permutation φ is given by the equation

$$\mathcal{L}(\varphi) = \prod_{i=1}^{n} h(x_i, y_{\varphi(i)}). \tag{2.4.5}$$

Assume that the prior probability of each permutation is 1/n!. Then the posterior probability that φ provides a completely correct set of n matches is

$$p(\varphi) = \mathcal{L}(\varphi) / \sum_{\psi \in \Phi} \mathcal{L}(\psi). \tag{2.4.6}$$

For $j = 1, 2, \ldots, n$, let

$$\Phi(j) = \{\varphi \in \Phi\colon \varphi(1) = j\} \tag{2.4.7}$$

be the set of $(n - 1)!$ permutations which specify that x_1 is to be paired with y_j. Using the definitions in (2.4.6) and (2.4.7), we get the posterior probability that the pairing of x_1 and y_j yields a correct match to be

$$p_j = \sum_{\varphi \in \Phi(j)} p(\varphi), \; 1 \leq j \leq n . \qquad (2.4.8)$$

For any two permutations φ and ψ in Φ, let $K(\varphi,\psi) = \# \{i : \varphi(i) = \psi(i)\}$ be the number of correct matches when the observations in the broken random sample are paired according to φ and the vectors in the original sample were actually paired according to ψ. It then follows that for any permutation $\varphi \in \Phi$, the quantity

$$M(\varphi) = \sum_{\psi \in \Phi} p(\psi) \, K(\varphi,\psi) \qquad (2.4.9)$$

is the posterior expected number of correct matches when φ is used to repair the data in the broken random sample. Finally, let $\Phi_{1,n}$ be the set of all permutations φ such that $y_{\varphi(1)}=y_1$ and $y_{\varphi(n)}=y_n$.

DeGroot et al (1971) have proposed three optimality criteria, for choosing the optimal matching strategy φ:

(i) maximize the posterior probability, $p(\varphi)$, of a completely correct set of n matches,

(ii) maximize the posterior probability, p_j, of correctly matching x_1 by choosing an optimal j from $\{1, 2,...., n\}$ and

(iii) maximize the posterior expected number, $M(\varphi)$, of correct matches in the repaired sample.

Assuming that the bivariate density of T and U was given by $h(t,u) = a(t)b(u)\, e^{tu}$, $(t,u) \in R^2$, the following results, among others, were established by them:

(a) The M.L.P. φ^* maximizes the probability of correct pairing of all n observations.

(b) The posterior probability of pairing $x_1(x_n)$ correctly is maximized by pairing $x_1(x_n)$ with $y_1(y_n)$.

(c) The class of permutations $\Phi_{1,n}$ is complete; that is, given any permutation $\varphi \notin \Phi_{1,n}$, there exists a $\psi \in \Phi_{1,n}$ which is as good as φ in the sense that $M(\psi) \geq M(\varphi)$.

(d) Sufficient conditions in terms of the data $x_1 < ... < x_n$ and $y_1 < ... < y_n$ for the M.L.P. φ^* to maximize $M(\varphi)$ were also given.

The results in Chew (1973) and Goel (1975) are extensions of (a) through (d) to an arbitrary bivariate density h(t,u) possessing the monotone likelihood ratio. The "completeness" property in (c) implies that the permutation φ^E maximizing $M(\varphi)$ satisfies $\varphi^E(1) = 1$ and $\varphi^E(n) = n$. It follows that for n = 2, 3, $\varphi^* \equiv \varphi^E$. DeGroot et al (1971) show that for n > 3, φ^E is not necessarily equal to the M.L.P. φ^* by means of a counter example.

2.4.4 Matching Problems for Multivariate Normal Distributions

In our review so far, we have discussed optimal matching strategies only in the case of bivariate data, one variable for each of the two files. However, multivariate data are often available in both files. Suppose that we have a model where (W) has a (p+q)- dimensional normal distribution with *known* variance-covariance matrix Σ. Let us write Σ and its inverse in the following partitioned form:

$$\Sigma = \begin{bmatrix} \Sigma_{11} & \Sigma_{12} \\ & \\ \Sigma_{21} & \Sigma_{22} \end{bmatrix} \quad \text{and } \Sigma^{-1} = \begin{bmatrix} \Omega_{11} & \Omega_{12} \\ & \\ \Omega_{21} & \Omega_{22} \end{bmatrix},$$

where both Σ_{12} and Ω_{12} have dimension p x q.

As before, we shall let $x_1, ..., x_n$ and $y_1, ..., y_n$ denote the values in a broken random sample from this distribution, where each x_i is a vector of dimension p x 1 and each y_j vector has the dimension q x 1. The results to be presented here were originally described by DeGroot and Goel (1976).

The likelihood function $\mathcal{L}$, as a function of the unknown permutation φ, can be written in the form

$$\mathcal{L}(\varphi) = \exp [- \frac{1}{2} \sum_{i=1}^{n} x_i' \, \tilde{\Omega}_{12} \, y_{\varphi(i)}], \qquad (2.4.10)$$

since the other factors in the joint density of the sample do not depend on φ. If we again assume that the prior probability of each permutation φ is $1/n!$, then the posterior probability that φ provides a completely correct set of n matches is given by (2.4.6). Thus, maximizing $p(\varphi)$ is equivalent to maximizing $\mathcal{L}(\varphi)$, or equivalently minimizing

$$Q(\varphi) = \sum_{i=1}^{n} \mathbf{x}_i' \, \Omega_{12} \, \mathbf{y}_{\varphi(i)} \qquad (2.4.11)$$

There is no simple way, in general, to describe the maximum likelihood solution.

However, if rank $(\Sigma_{12}) = 1$, then rank $(\Omega_{12}) = 1$ and Ω_{12} can be represented in the form $\Omega_{12} = \mathbf{a}'\mathbf{b}$, where $\mathbf{a}$ and $\mathbf{b}$ are vectors of dimensions p x 1 and q x 1. If we let $\gamma(\mathbf{x}_i) = \mathbf{a}'\mathbf{x}_i$ and $\delta(\mathbf{y}_i) = \mathbf{b}'\mathbf{y}_i$ for i = 1, 2,..., n, then φ^* will be the permutation that minimizes

$$Q(\varphi) = \sum_{i=1}^{n} \gamma(\mathbf{x}_i) \; \delta(\mathbf{y}_{\varphi(i)}). \qquad (2.4.12)$$

Now, minimizing (2.4.12) is achieved by arranging $\gamma(x_i)$'s from smallest to largest, arranging $\delta(y_j)$'s in the reverse order from the largest to smallest and then pairing the corresponding elements in the two sequences.

Suppose next that rank $(\Omega_{12}) \geq 2$. Without loss of generality, we shall assume that $p \leq q$ and let $\mathbf{y}_j^* = \Omega_{12}\mathbf{y}_j$, for j = 1, 2,..., n. Then both $\mathbf{x}_i$ and $\mathbf{y}_j^*$ are p-dimensional vectors, and the maximum likelihood solution φ^* will be the permutation that minimizes

$$\underset{\sim}{Q}(\varphi) = \sum_{i=1}^{n} \mathbf{x}_i' \, \mathbf{y}^*_{\varphi(i)}.$$

Let D denote the n x n matrix, whose elements are $d_{ij} = \mathbf{x}_i' \, \mathbf{y}_j$. Then minimizing $\underset{\sim}{Q}(\varphi)$ is equivalent to minimizing

$$\sum_{i=1}^{n} \sum_{j=1}^{n} d_{ij}\, a_{ij} \tag{2.4.13}$$

subject to the constraints

$$\sum_{i=1}^{n} a_{ij} = 1, \text{ for } j = 1, 2, \ldots, n,$$

$$\sum_{j=1}^{n} a_{ij} = 1, \text{ for } i = 1, 2, \ldots, n, \tag{2.4.14}$$

$$a_{ij} = 0 \text{ or } 1,$$

which is a standard assignment problem with cost matrix **D**. Although, there is no simple form for the solution of an arbitrary assignment problem of this type, efficient algorithms are available for finding numerical solutions.

When p and n are moderately large, the permutation φ^E, that maximizes the expected number of correct matches, is very difficult to calculate. No efficient algorithms are known. A Monte-Carlo study was reported by DeGroot and Goel (1976) in which they compare φ^E and φ^* for $p = 2$ and 50 different covariance matrices Σ with the sample size $n = 3, 4$ and 5. In all cases, the proportion of samples for which φ^E and φ^* were identical is between 0.925 and 0.995. Thus, it is not unreasonable to use φ^* even when the goal is to maximize the posterior expected number of correct matches.

DeGroot and Goel (1976) studied two other simple matching strategies which provide good approximations to the M.L.P. φ^* or to the rule φ^E. In the rest of this chapter, we shall discuss matching problems only in the bivariate case.

2.5. Reliability of Matching Strategies for Bivariate Data

Consider a random sample $W_1, \ldots, W_n$, from a bivariate population with density h. Even if the pairings in this sample are lost before the entire data was recorded, we can observe the marginal order-statistics. In fact, if $X_1, \ldots, X_n$ and $Y_1, \ldots, Y_n$ is the broken random sample corresponding to the unobserved sample on the pair **W**, then clearly the order-statistics $X_{(1)} < \ldots < X_{(n)}$ of the x's are exactly the same as the order-statistics $T_{(1)} < \ldots < T_{(n)}$ of the T's. Similarly, the order-statistics $Y_{(1)} < \ldots < Y_{(n)}$ are the same as $U_{(1)} < \ldots < U_{(n)}$. In repairing of the x's and y's, introduced in Section 2.4, each permutation φ in Φ provides a matching strategy such that typical merged file consists of the pairs

$$(X_{(i)},\ Y_{(\varphi(i))}),\ i = 1, 2, \ldots, n, \tag{2.5.1}$$

We are now concerned with the quality of this synthetic file.

Ideally, we would like to choose a φ so that the file in (2.5.1) recovers all the unobserved pairs. It is, therefore, natural to study the properties of the random variable $N(\varphi)$, *the number of correct matches due to* φ or, equivalently, the number of unobserved sample points which have been recovered in (2.5.1). It should be pointed out that $N(\varphi)$ is one of the two measures of reliability of the strategy φ suggested in Section 1.6, .

2.6 An Optimality Property of the Matching Strategy φ^*

Some optimality properties of the matching strategy φ^* (the M.L.P. defined in Section 2.4.2) from a Bayesian point of view, were reviewed in Section 2.4.3. We shall now establish optimality of φ^* from the frequentist view point.

Consider the random variable $N(\varphi)$, the number of correct matches resulting from the use of the permutation φ in Φ to merge the broken random sample from a bivariate population. It should be pointed out that $M(\varphi)$, which was defined in Section 2.4.3, is different from $E(N(\varphi))$ because the former quantity is a posterior expected value given a particular broken random sample and, in the latter, the expectation is taken over all possible samples.

Remark: The number of correct matches resulting from the matching strategy φ has the useful representation

$$N(\varphi) = \sum_{i=1}^{n} I_{(S_i = \varphi(R_i))}. \tag{2.6.1}$$

In this section, we shall show that φ^* maximizes $E(N(\varphi))$, the expected number of correct matches, provided that the parent density $h(t,u)$ exhibits certain dependence structures described below.

Definition 2.6.1. (Shaked, 1979) *Exchangeable random variables T,U are said to be* ***positive dependent by mixture*** (PDM) *iff the joint distribution of T, U is that of* $g(\underset{\sim}{\xi}_0,\xi_1)$ *and* $g(\underset{\sim}{\xi}_0,\xi_2)$, *where* ξ_1 *and* ξ_2 *are i.i.d. random variables,* $\underset{\sim}{\xi}_0$ *is a random vector which is independent of* ξ_1 *and* ξ_2 *and* g *is a Borel measurable function.*

Definition 2.6.2. (Shaked, 1979) *Exchangeable random variables T,U are said to be* ***positive dependent by expansion*** (PDE) *iff the joint distribution of T and U admits the following series expansion:*

$$dH(t,u) = [1 + \sum a_i \, \eta_i(t) \, \eta_i(u)] \, dF(t) \, dF(u) \qquad (2.6.2)$$

where $F(\cdot)$ *is the marginal CDF of T or U,* a_i*'s are nonnegative real numbers, and* $\{\eta_i\}$ *is a set of functions satisfying*

$$\int_{-\infty}^{\infty} \eta_i(x) \, dF(x) = 0, \; i = 1, 2, \ldots . \qquad (2.6.3)$$

According to the Definitions 2.6.1 and 2.6.2, the dependence concepts will apply only to pairs of exchangeable random variables. It may also be noted that for most of the known expansions of PDE distributions, the set of functions $\{\eta_k(\cdot)\}$ satisfies, in addition to (2.6.3), the orthogonality conditions

$$\int_{-\infty}^{\infty} \eta_k(x) \, \eta_m(x) \, dF(x) = \delta_{km} , \qquad (2.6.4)$$

where $k, m = 1, 2, \ldots,$ and δ_{km} is the Kronecker delta.

We now give two examples to illustrate these concepts of dependence.

Example 2.6.1. Let ξ_0, ξ_1, ξ_2 be i.i.d. standard normal random variables. Let ρ be any constant in the interval [0,1]. Define new random variables

$$T = \sqrt{1-\rho} \cdot \xi_1 + \sqrt{\rho}\ \xi_0$$

$$U = \sqrt{1-\rho} \cdot \xi_2 + \sqrt{\rho}\ \xi_0$$

Then, it is easy to verify that T,U are jointly normal and that the definition (2.6.1) can be applied to T and U with the above choice of ξ_0, ξ_1, ξ_2. Hence, the standard bivariate normal distribution with nonnegative correlation has the PDM property.

Also, Mardia (1970, p. 48) gives the following series expansion for the bivariate normal density

$$h(t,u) = [\ 1 + \sum_{k=1}^{\infty} \rho^k \eta_k(t)\eta_k(u)\ ]\ f(t)\ f(u), \tag{2.6.5}$$

where f(t) is the density of the univariate standard normal random variable and $\{\eta_k(\cdot)\}$ is a set of orthonormal Hermite polynomials. Thus, if $\rho \geq 0$, bivariate normal distributions possess the PDE property as well.

Example 2.6.2. A class of bivariate densities, known as Farlie-Gumbel-Morgenstern distribution, is given by the formula

$$h(t,u) = 1 + \alpha(1 - 2t)(1 - 2u), \text{ where } 0 < t, u < 1 \text{ and } -1 \leq \alpha \leq 1 \tag{2.6.6}$$

It is easy to check that T and U are PDE for $\alpha \geq 0$ in (2.6.6). Note that the expansion (2.6.6) has only a finite number of terms, unlike the expansion for the bivariate normal distribution.

The following result on the exchangeability of random variables (see Randles and Wolfe (1979)) will be needed in the sequel.

Theorem 2.6.1. *If* $\underset{\sim}{\xi} \overset{d}{=} \underset{\sim}{\eta}$ *and* $\mathbf{K}(\cdot)$ *is a measurable function (possibly vector valued) defined on the common support of these random vectors, then* $\mathbf{K}(\underset{\sim}{\xi}) \overset{d}{=} \mathbf{K}(\underset{\sim}{\eta})$.

We now prove that the PDM/PDE structures are inherited by a pair of new variables obtained from a given sample by computing the *same* function of the marginals. These results are generalizations of theorems in Shaked (1979), which were proved only for n=2, and are needed for establishing the optimality property of φ^*.

Theorem 2.6.2. *Let* $\mathbf{W}_i$, $i = 1, 2, \ldots, n$ *be a random sample from a bivariate PDM parent with density* h(t,u). *Then, for any measurable function* $g: R^n \to R$, *the random variables* $g(T_1, T_2, \ldots, T_n)$ *and* $g(U_1, U_2, \ldots, U_n)$ *are jointly PDM.*

Proof. By hypothesis, the vectors $\mathbf{W}_i$ are i.i.d. Furthermore, since PDM property is defined only for exchangeable pairs of random variables, we have

$$(T_i, U_i) \overset{d}{=} (U_i, T_i), \; i = 1, 2, \ldots, n. \tag{2.6.7}$$

Equation (2.6.7) together with the independence of (T,U) pairs yields

$$(T_1, \ldots, T_n, U_1, \ldots, U_n) \overset{d}{=} (U_1, \ldots, U_n, T_1, \ldots, T_n) \tag{2.6.8}$$

Consider the function $\mathbf{K}: R^{2n} \to R^n$, defined by the equation

$$\mathbf{K}(a_1, \ldots, a_n; b_1, \ldots, b_n) = (g(a_1, \ldots, a_n), g(b_1, \ldots, b_n))$$

where $(a_1, \ldots, a_n, b_1, \ldots, b_n)$ is any point in R^{2n}. Applying the function $\mathbf{K}$ to both sides of (2.6.8) and invoking Theorem 2.6.1 we get

$$(g(T_1, \ldots, T_n), g(U_1, \ldots, U_n)) \overset{d}{=} (g(U_1, \ldots, U_n), g(T_1, \ldots, T_n)) \tag{2.6.9}$$

Hence, (g(**T**), g(**U**)) is an exchangeable-pair of random variables.

The PDM property of (T_i, U_i), $i = 1, 2, \ldots, n$ further implies that there exist n i.i.d. vectors $(\underset{\sim}{\xi}_{0i}, \xi_{1i}, \xi_{2i})$, $i = 1, 2, \ldots, n$ and a measurable function f such that

(i) For each j, ξ_{1j}, ξ_{2j} are i.i.d. univariate random variables and the vector $\underset{\sim}{\xi}_{0j}$ is independent of ξ_{1j} and ξ_{2j}.

(ii) For each j,

$$T_j = f(\xi_{1j}, \underset{\sim}{\xi}_{0j}) \text{ and } U_j = f(\xi_{2j}, \underset{\sim}{\xi}_{0j}). \tag{2.6.10}$$

Introducing the random variables,

$$\xi_1^* = \xi_{11}, \quad \xi_2^* = \xi_{21}$$

and

$$\underset{\sim}{\xi}_0^* = (\xi_{12},\ldots, \xi_{1n}, \xi_{22},\ldots, \xi_{2n}, \underset{\sim}{\xi}_{01},\ldots, \underset{\sim}{\xi}_{0n}), \qquad (2.6.11)$$

we find that ξ_1^* and ξ_2^* are i.i.d. univariate random variables and $\underset{\sim}{\xi}_0^*$ is independent of ξ_1^* and ξ_2^* in view of the properties (i) and (ii). Note that (2.6.10) and (2.6.11) imply that

$$g(T) = g(f(\xi_{11}, \underset{\sim}{\xi}_{01}), \ldots, f(\xi_{1n}, \underset{\sim}{\xi}_{0n}))$$

is a measurable function g^*, say, of ξ_1^* and $\underset{\sim}{\xi}_0^*$. Similarly, g(U) is also the *same* function g^* of the random variables ξ_2^* and $\underset{\sim}{\xi}_0^*$. Hence, by definition, g(T) and g (U) are PDM.

The next result is similar to Theorem 2.6.2, when the parent distribution has the PDE property.

Theorem 2.6.3. *Let* W_i, *i = 1,..., n be a random sample from a PDE parent. Then, for any measurable function* $g: R^n \to R$, *the random variables* $g(T_1,\ldots,T_n)$ *and* $g(U_1,\ldots, U_n)$ *are PDE.*

Proof. The exchangeability of the joint distribution of g(T) and g(U) has already been proved in Theorem 2.6.2 (see equation 2.6.9). It remains to be shown that, when the joint density of each of the n copies of T,U admits an expansion in (2.6.2), the joint density of g(T) and g(U) also admits a similar expansion.

Assume therefore that there exists nonnegative constants $\{a_k\}$ and a set of functions $\{\eta_k(\cdot)\}$ satisfying (2.6.3) such that the joint density of T_i and U_i is of the form

$$dH(t_i,u_i) = dF(t_i)dF(u_i)\,[1 + \sum_{k=1}^{\infty} a_k\, \eta_k(t_i)\, \eta_k(u_i)],\ i = 1, 2,\ldots, n. \qquad (2.6.12)$$

For any real x, define the measurable set in R^n

$$A(x) = \{\ (x_1,\ldots, x_n):\ g(x_1,\ldots, x_n) \le x\ \}.$$

Then, the distribution function Q, of (g(T), g(U)) is

$$Q(x,y) = \int_{t \in A(x)} \cdots \int \int_{u \in A(y)} \cdots \int \prod_{j=1}^{n} dH(t_j,u_j) \tag{2.6.13}$$

Using the expansions in equation (2.6.12) we get

$$\begin{aligned} Q(x,y) = \tilde{Q}(x)\,\tilde{Q}(y) &+ n \sum_{k=1}^{\infty} a_k \chi_k^{(1)}(x)\chi_k^{(1)}(y) \\ &+ \binom{n}{2} \sum_{k=1}^{\infty} \sum_{\ell=1}^{\infty} a_k a_\ell \chi_{k,\ell}^{(2)}(x)\chi_{k,\ell}^{(2)}(y) + \ldots \\ &+ \sum_{k_1=1}^{\infty} \cdots \sum_{k_n=1}^{\infty} a_{k_1} \cdots a_{k_n} \chi_{k_1,\ldots,k_n}^{(n)}(x)\chi_{k_1,\ldots,k_n}^{(n)}(y), \end{aligned} \tag{2.6.14}$$

where

$$\tilde{Q}(x) = \int_{A(x)} \cdots \int \prod_{i=1}^{n} dF(t_i),$$

$$\chi_k^{(1)}(x) = \int_{A(x)} \cdots \int \eta_k(t_1) \prod_{i=1}^{n} dF(t_i),$$

$$\chi_{k,\ell}^{(2)}(x) = \int_{A(x)} \cdots \int \eta_k(t_1)\,\eta_\ell(t_2) \prod_{i=1}^{n} dF(t_i),$$

and

$$\chi_{k_1,\ldots,k_n}^{(n)}(x) = \int_{A(x)} \cdots \int \prod_{i=1}^{n} \eta_{k_i}(t_1) \prod_{i=1}^{n} dF(t_i). \tag{2.6.15}$$

Note that $\forall\, k_i = 1, 2, \ldots$ and $\forall\, i = 1, 2, \ldots, n$ the signed measure induced by $\chi_{k_1,\ldots,k_\ell}^{(\ell)}(x)$ is absolutely continuous with respect to $\tilde{Q}$ so that there exists $\psi_{k_1,\ldots,k_\ell}^{(\ell)}(x)$ - the Radon-Nikodym derivative - of $\chi^{(\ell)}(x)$ with respect to $\tilde{Q}$ such that

$$\chi^{(\ell)}_{k_1,\ldots,k_\ell}(x) = \int_{-\infty}^{x} \psi^{(\ell)}_{k_1,\ldots,k_\ell}(t)\, d\tilde{Q}(t)\,. \tag{2.6.16}$$

Hence, from equations (2.6.14) to (2.6.16) we get

$$\begin{aligned} dQ(x,y) = d\tilde{Q}(x)\, d\tilde{Q}(y)\, [1 &+ n \sum_{k=1}^{\infty} a_k \psi_k^{(1)}(x)\, \psi_k^{(1)}(y) \\ &+ \binom{n}{2} \sum_{k_1=1}^{\infty} \sum_{k_2=1}^{\infty} a_{k_1} a_{k_2} \psi^{(2)}_{k_1,k_2}(x)\, \psi^{(2)}_{k_1,k_2}(y) + \ldots \\ &+ \sum_{k_1=1}^{\infty} \cdots \sum_{k_n=1}^{\infty} a_{k_1} \cdots a_{k_2}\, \psi^{(n)}_{k_1,\ldots,k_n}(x)\, \psi^{(n)}_{k_1,\ldots,k_n}(y). \end{aligned} \tag{2.6.17}$$

The representation (2.6.17) holds almost everywhere ($\tilde{Q}$ measure) because Radon-Nikodym derivatives are defined up to sets of measure zero. Also, the coefficients in (2.6.17), being products of the nonnegative a_k's, are themselves nonnegative. To complete the proof, we only have to show that the orthogonality conditions (2.6.3) hold for the Ψ_k's in the expansion (2.6.17).

For $\ell = 1, 2, \ldots, n$, and $1 \le k_1, \ldots, k_\ell < \infty$ we have

$$\begin{aligned} \int_{-\infty}^{\infty} \psi^{(\ell)}_{k_1,\ldots,k_\ell}(t)\, d\tilde{Q}(t) &= \lim_{x\to+\infty} \chi^{(\ell)}_{k_1,\ldots,k_\ell}(x) \\ &= \int_{-\infty}^{\infty} \cdots \int_{-\infty}^{\infty} \prod_{i=1}^{\ell} \eta_{k_i}(t_i) \prod_{i=1}^{n} dF(t_i) \\ &= \Big[\int_{-\infty}^{\infty} \eta_{k_1}(t_1)\, dF(t_1) \Big] \Big[\int_{-\infty}^{\infty} \cdots \int_{-\infty}^{\infty} \prod_{i=2}^{\ell} \eta_{k_i}(t_i) \prod_{i=2}^{n} dF(t_i) \Big]. \end{aligned} \tag{2.6.18}$$

By hypothesis $\{\eta_k(\cdot)\}$ satisfy (2.6.3). Therefore the first term in (2.6.18) is zero. Hence,

$$\int_{-\infty}^{\infty} \psi^{(\ell)}(t)\, d\tilde{Q}(t) = 0, \text{ where } \ell = 1, 2, \ldots \tag{2.6.19}$$

and this completes the proof.

The following facts about bivariate ranks are easy consequences of Theorem 2.6.2 and Theorem 2.6.3.

Corollary 2.6.1. *Let* $\mathbf{W}_i$, *i = 1, 2, ..., n be a random sample from a PDM (PDE) parent. Consider the marginal ranks* R_i, S_i *of* T_i *and* U_i *respectively, defined in Section 2.2. The pair* (R_i, S_i) *is PDM (PDE),* i = 1, 2,..., n.

Proof. Fix i and define a function g: $R^n \to R$ by the equation

$$g_i(a_1,..., a_n) = \sum_{\alpha=1}^{n} I_{(a_i \geq a_\alpha)}$$

and observe that $R_i = g_i(T_1,...,T_n)$, $S_i = g_i(U_1,...,U_n)$. By invoking Theorems 2.6.2 and 2.6.3, the result follows.

Following result will be needed for establishing an optimality property of φ^*.

Theorem 2.6.4. *Let random vectors* $\mathbf{W}_i$, *i = 1, 2,..., n, be PDM/PDE. Let* R_1, S_1 *denote the ranks of* T_1, U_1, *among* T_i's *and* U_j's, *respectively. Consider the joint probability mass function*

$$\pi_{ij} = P(R_1 = i, S_1 = j), \ 1 \leq i, j \leq n$$

of R_1 *and* S_1. *Then* $\forall$ i, j, π_{ij}'s *satisfy the following inequalities:*

$$\pi_{ii} + \pi_{jj} \geq 2\pi_{ij} \tag{2.6.20}$$

Proof. By hypothesis, the parent distribution is PDM (PDE). It follows from Corollary 2.6.1, that R_1 and S_1 are also PDM (PDE), exchangeable random variables. Hence,

$$\pi_{ij} = \pi_{ji}, \ \text{for } 1 \leq i, j \leq n \tag{2.6.21}$$

To establish (2.6.20), first consider the case when T and U are PDM, and consequently R_1 and S_1, are PDM. There exists a distribution function Q(·) of some d-dimensional random vector

$$\pi_{ij} = \int_{-\infty}^{\infty} \cdots \int_{-\infty}^{\infty} \pi_i(t) \ \pi_j(t) \ dQ(t), \ 1 \leq i, j \leq n \tag{2.6.22}$$

where $\pi_i(t)$ and $\pi_j(t)$ are the conditional mass functions of R_1 and S_1, given a value **t** from the Q-distribution. It follows from equation (2.6.22) that

$$\pi_{ii} + \pi_{jj} - 2\pi_{ij}$$

$$= \int_{-\infty}^{\infty} \cdots \int_{-\infty}^{\infty} [(\pi_i(t))^2 + (\pi_j(t))^2 - 2\pi_i(t)\,\pi_j(t)]\, dQ(t)$$

$$= \int_{-\infty}^{\infty} \cdots \int_{-\infty}^{\infty} (\pi_i(t) - \pi_j(t))^2\, dQ(t) \geq 0 .$$

We thus obtain (2.6.20) when T, U are PDM.

Suppose now that T and U are PDE. Then, by virtue of Corollary 2.6.1, R_1 and S_1 would be PDE. Furthermore, R_1 and S_1 are both discrete uniform random variables on 1, 2,..., n [see Randles and Wolfe (1979), p. 38].

As R_1 and S_1 have finite supports the series expansion of R_1 and S_1 will have a finite number of terms. In fact, Fisher's identity (see Lancaster (1969), p. 90) holds i.e.,

$$\pi_{ij} = \frac{1}{n}\cdot\frac{1}{n}(1 + \sum_{k=1}^{n-1} a_k\eta_k(i)\,\eta_k(j)),\ 1 \leq i, j \leq n, \qquad (2.6.23)$$

where $\{a_k\}$ are nonnegative constants and $\{\eta_k(\cdot)\}$ are orthogonal functions on 1, 2,..., n. The representation (2.6.23) implies that for $1 \leq i, j \leq n$,

$$\pi_{ii} + \pi_{jj} - 2\pi_{ij} = \frac{1}{n^2}[\sum_{k=1}^{n-1} a_k(\eta_k(i))^2 + \sum_{k=1}^{n-1} a_k(\eta_k(j))^2 - 2\sum_{k=1}^{n-1} a_k\eta_k(i)\,\eta_k(j)]$$

$$= \frac{1}{n^2}\sum_{k=1}^{n-1} a_k[\eta_k(i) - \eta_k(j)]^2] \geq 0 \qquad (2.6.24)$$

Hence, we obtain the inequalities in (2.6.20). The optimality property of φ^* can now be established:

Theorem 2.6.5. *Let* $\mathbf{W}_i$, $i = 1, 2, \ldots, n$ *be a sample from a PDM/PDE distribution. Then*

$$E[N(\varphi^*)] \geq E[N(\varphi)] \text{ for all } \varphi \in \Phi, \tag{2.6.25}$$

i.e., φ^* *is an optimal strategy for maximizing the expected number of correct matches,* $E(N(\varphi))$.

Proof. Since $N(\varphi)$ is the number of correct matches, using equation 2.6.1, we can write

$$E(N(\varphi)) = nP(S_1 = \varphi(R_1)) \tag{2.6.26}$$

$$= n \sum_{k=1}^{n} P(S_1 = \varphi(k), R_1 = k)$$

$$= n \sum_{k=1}^{n} \pi_{k,\varphi(k)} \, ,$$

where π is the joint mass function of R_1, S_1. Invoking the inequalities for π_{ij} given by (2.6.20) we obtain

$$E(N(\varphi)) \leq n \sum_{k=1}^{n} \frac{1}{2} (\pi_{k,k} + \pi_{\varphi(k),\varphi(k)})$$

$$= n \left[\frac{1}{2} \sum_{k=1}^{n} \pi_{k,k} + \frac{1}{2} \sum_{k=1}^{n} \pi_{\varphi(k),\varphi(k)}\right]$$

$$= n \sum_{i=1}^{n} \pi_{i,i} = n\,P(S_1 = R_1) = E(N(\varphi^*)).$$

This establishes the desired result.

Shaked (1979) has shown that a PDM/PDE parent need not have the MLR property and vice-versa, although some distributions such as the bivariate normal do possess both PDM(or PDE) and MLR structures.

Remark. *In view of Theorem 2.6.5 for PDM(PDE) families, we conjecture that* E(N(φ)) *is maximum at* φ = φ* *when the parent density has the MLR property.*

2.7. Monotonicity of E(N(φ*)) with respect to Dependence Parameters

Repairing of broken random samples based on the available data in two files was discussed in Section 2.4, where model based optimal matching strategies were derived for data coming from populations having certain types of positive dependent structures. Intuitively, it is reasonable to expect that the performance of the optimal matching strategy increases as the degree of positive dependence in the population increases. Our objective in this section is to present this type of results for the pairing φ*. We begin with a definition from Shaked (1979):

Definition 2.7.1. *Let J be a subset of* R. *A kernel* K *defined on J×J is said to be* ***conditionally positive definite*** (cpd) *on J×J iff*

(i) $K(x,y) = K(y,x)$, $\forall\, x,y \in J$; that is K is a symmetric kernel.

(ii) Let m be any positive integer. For arbitrary real numbers $a_1,\dots, a_m$ and for every choice of distinct numbers $x_1,\dots, x_m$ from J,

$$\sum_{i=1}^{m}\sum_{j=1}^{m} K(x_i,x_j)\, a_i\, a_j \geq 0 \text{ whenever } \sum_{i=1}^{m} a_i = 0\,. \tag{2.7.1}$$

It is pertinent to note that cpd kernels are a special case of positive definite kernels, because the nonnegativity of the quadratic form in (2.7.1), without requiring that the A_i's sum to zero, is a standard definition of positive definite kernels (Widder, 1941, p. 271). We shall now give an example of a cpd kernel which will also be used in the sequel.

Example 2.7.1. Let J = {1, 2,..., n}, where n is a fixed positive integer. Then the kernel $g(x,y) = I_{[x=y]}$ is positive definite on JxJ. For arbitrary real numbers $a_1,\dots, a_m$ and for every choice of distinct integers $i_1,\dots, i_m$ from J, we have

$$\sum_{\alpha=1}^{m}\sum_{\beta=1}^{m} g(i_\alpha, i_\beta)\ a_\alpha a_\beta = \sum_{\alpha=1}^{m} a_\alpha^2. \tag{2.7.2}$$

Here we use the fact that $i_\alpha = i_\beta$ iff $\alpha = \beta$, since integers $i_1, \ldots, i_m$ are distinct. The function $I_{(x=y)}$ is clearly symmetric in x and y. Hence, it follows from (2.7.2) that g(x,y) is positive definite.

The following lemma will be required to prove an optimality property of $N(\varphi^*)$ in Theorem 2.7.1.

Lemma 2.7.1 (Shaked, 1979). *Let* T *and* U *be PDM or PDE random variables with joint distribution function* H(t,u). *Let* $H_o(t,u)$ *be the distribution function in the case when* T *and* U *are independently distributed. Then*

$$E_H(K(T,U)) \geq E_{H_o}(K(T,U)) \qquad (2.7.3)$$

iff K(.,.) is a cpd kernel, provided the expectations exist.

Theorem 2.7.1. *Let* T *and* U *be PDM(PDE) random variables with joint distribution function* H(t,u). *If* $N \equiv N(\varphi^*)$ *is the number of correct matches due to the matching strategy* φ^*, *then*

$$E_H(N) \geq 1. \qquad (2.7.4)$$

Proof. It follows from equation (2.6.26) that

$$E_H(N) = n\, P_H(R_1 = S_1) = n\, E_H(g(R_1, S_1)) \qquad (2.7.5)$$

where $g(x,y) = I_{(x=y)}$. Now, recall from Example 2.7.1 that g(x,y) is p.d. on the domain JxJ, where J = {1, 2,..., n} is the common support of R_1 and S_1. It was established in Corollary 2.6.1 that R_1 and S_1 are PDM (PDE) according as T and U are PDM (PDE). Invoking Lemma 2.7.1, we therefore obtain

$$E_H[g(R_1, S_1)] \geq E_{H_o}[g(R_1, S_1)]. \qquad (2.7.6)$$

Equations (2.7.5) and (2.7.6) imply that

$$E_H(N) \geq E_{H_o}(N) \qquad (2.7.7)$$

Under H_o, T and U are independent random variables. Therefore, as discussed in Section 2.3, $E_{H_o}[N] = 1$. We conclude from (2.7.4) that φ^* provides, on the average, more correct matches when the data in the two files come from certain positively dependent populations than when they are independent. In particular, this fact holds for the bivariate normal distribution with positive correlation as well as for Morgenstern distributions in equation (2.6.6), when the dependence parameter $\alpha \geq 0$. Because of the inequality (2.7.7), a natural conjecture is that $E_H(N)$, as a functional of the distribution function H, is order-preserving with regard to certain partial orderings of the space of all continuous bivariate distributions with fixed marginals (of T and U) which exhibit positive dependence. Although no proof of this conjecture is available at this time, we offer further evidence in support of this conjecture in the next two theorems.

Theorem 2.7.2. *Suppose that a broken random sample comes from the family of densities given by the equation*

$$h(t,u) = 1 + \alpha(1-2t)(1-2u),\ 0 < t, u < 1 \text{ and } 0 \leq \alpha < 1 \qquad (2.7.8)$$

Then, $E_\alpha(N)$ *is monotone increasing in* α.

Proof. Note that in (2.7.8), $\alpha = 0$ means T and U are independent and we might say that the farther α is from 0 the more the positive dependence between T and U.

It follows from equation (2.6.17) and Corollary 2.6.1 that the joint probability function of the ranks R_1 and S_1 can be canonically expanded as follows:

$$\pi_{ij} = P(R_1 = i, S_1 = j)$$

$$= \frac{1}{n^2}\left[1 + \sum_{k=1}^{n} \binom{n}{k} \alpha^k \eta_k(i)\, \eta_k(j)\right] \qquad (2.7.9)$$

where i, j = 1, 2,..., n and $\{\eta_k(\cdot)\}_1^n$ is a set of functions satisfying the orthogonality conditions in (2.6.3). Using the expression (2.7.9) for π_{ij} we get

$$E_\alpha(N) = n\,P(R_1 = S_1)$$

$$= n \sum_{i=1}^{n} \pi_{ii}$$

$$= n \cdot \frac{1}{n^2} [n + \sum_{i=1}^{n} \sum_{k=1}^{n} \binom{n}{k} \alpha^k (\eta_k(i))^2]$$

$$= 1 + \frac{1}{n} \sum_{k=1}^{n} \binom{n}{k} b_k \alpha^k \qquad (2.7.10)$$

where, after a change in the order of summations on i and k, we have used nonnegative constants b_k given by the equation

$$b_k = \sum_{i=1}^{n} (\eta_k(i))^2, \; k = 1, 2, \ldots, n .$$

It follows from (2.7.10) that $E_\alpha(N)$ is a polynomial in α with non-negative coefficients. Hence it increases with α, as α goes from 0 to 1.

Theorem 2.7.3. *Suppose that a broken random sample comes from the bivariate normal distributions given by (2.6.5), where we assume that the correlation parameter* ρ *is nonnegative. Then* $E_\rho(N)$ *is increasing in* ρ.

Proof. It follows from equation (2.6.17) and Corollary 2.6.1 that

$$\pi_{ij} = P(R_1 = i, S_1 = j)$$

$$= \frac{1}{n^2} [1 + n \sum_{k=1}^{\infty} \rho \, \psi_k^{(1)}(i) \, \psi_k^{(1)}(j)$$

$$+ \binom{n}{2} \sum_{k_1, k_2 = 1}^{\infty} \rho^2 \, \psi_{k_1, k_2}^{(2)}(i) \, \psi_{k_1, k_2}^{(2)}(j) + \ldots$$

$$+ \sum_{k_1, \ldots, k_n = 1}^{\infty} \rho^n \, \psi_{k_1, \ldots, k_n}^{(n)}(i) \, \psi_{k_1, \ldots, k_n}^{(n)}(j)]. \qquad (2.7.11)$$

where, for fixed $\ell = 1, 2, \ldots,$ $\{\psi_{k_1, \ldots, k_\ell}^{(\ell)}\}$ is a set of functions on $\{1, 2, \ldots, \ell\}$ satisfying (2.6.3). Using the expression (2.7.11) for π_{ii}, we obtain

$$E_\rho(N) = nP(R_1 = S_1)$$

$$= n \sum_{i=1}^{n} \pi_{ii}$$

$$= \frac{1}{n} [n + n \cdot \rho \sum_{k=1}^{\infty} \sum_{i=1}^{n} (\psi_k^{(1)}(i))^2$$

$$+ \binom{n}{2} \rho^2 \sum_{k_1,k_2=1}^{\infty} \sum_{i=1}^{n} (\psi_{k_1,k_2}^{(2)}(i))^2$$

$$+ \ldots + \rho^n \sum_{k_1,\ldots,k_n=1}^{\infty} \sum_{i=1}^{n} (\psi_{k_1,\ldots,k_n}^{(n)}(i))^2], \qquad (2.7.12)$$

where the order of summations over i and $k_1,\ldots, k_n$ have been reversed because the terms in the expansion for π_{ii} given by (2.7.11) are all nonnegative. We conclude from (2.7.12) that $E_\rho(N)$ is a Power-series in ρ with non-negative coefficients. Hence it increases with ρ as ρ goes from 0 to 1.

As we close this section, we shall state a result due to Chew (1973) which somewhat resembles, though conceptually different from, the inequality $E_H(N) \geq 1$ in (2.7.4). Recall the notation $M(\varphi)$ in (2.4.9), which denotes the posterior expected number of correct matches due to the strategy φ. Arguing that $M(\varphi) = 1$ when φ is randomly chosen from Φ, he proved the following result:

Theorem 2.7.4. (Chew, 1973). *Let* $x_1,\ldots, x_n$ *and* $y_1,\ldots, y_n$ *be a broken random sample from a bivariate distribution possessing monotone likelihood ratio. The posterior expected number of correct pairings using the M.L.P.* φ^* *satisfies*

$$M(\varphi^*) \geq 1. \qquad (2.7.13)$$

It should be noted that the inequality (2.7.13) was derived from a Bayesian perspective. Whereas, in our inequality (2.7.4), the expectation is over all possible samples.

2.8. Some Properties of Approximate Matching Strategies

In practice, situations often arise in which the requirement that the matched pairs in the merged file be exactly the same as the pairs of the original data is not crucial. For example, when contingency tables analyses are contemplated for grouped data on continuous variables T and U then, in the absence of the knowledge of the pairings, we would like to reconstruct the pairs but would not worry too much as long as the u-value in any matched pair came within a pre-fixed tolerance ε ($\varepsilon > 0$) of the true u-value that we would get with the ideal matching which recovers all the original pairs. This type of 'approximate matching' was introduced by Yahav (1982), who defined ε -correct matching as follows.

Definition 2.8.1 (Yahav). *In the merged file* (2.5.1), *a pair* $(T_{(i)}, U_{(\varphi(i))})$ *is said to be* ε *-correct, if* $| U_{(\varphi(i))} - U_{[i]} | \leq \varepsilon$, *here* $U_{[i]}$ *is the concomitant of* $T_{(i)}$; *that is the true u-value that was paired with* $T_{(i)}$ *in the original sample.*

The number of ε -correct matches $N(\varphi, \varepsilon)$, in the merged file (2.5.1) can be written as

$$N(\varphi,\varepsilon) = \sum_{i=1}^{n} I\,[\,|\,U_{(\varphi(i))} - U_{[i]}\,| \leq \varepsilon\,] \tag{2.8.1}$$

Note that $N(\varphi)$, the number of correct matches due to φ, can be obtained from $N(\varphi,\varepsilon)$ by formally letting $\varepsilon = 0$.

The counts $N(\varphi)$ and $N(\varphi,\varepsilon)$ are useful indices reflecting the reliability of the merged file resulting from φ. Sections 2.6 and 2.7 dealt with a few properties of $N(\varphi^*)$. Some statistical properties of $N(\varphi^*,\varepsilon)$ given here have also been discussed in Goel and Ramalingam (1987).

The following representation for $N(\varphi,\varepsilon)$ as a sum of exchangeable 0 -1 random variables will be useful for extending results of Yahav (1982).

Theorem 2.8.1. *Let* $N(\varphi,\varepsilon)$ *be as defined in* (2.8.1). *Then, for all* $\varphi \in \Phi$,

$$N(\varphi,\varepsilon) = \sum_{i=1}^{n} \upsilon_{ni}(\varphi,\varepsilon), \tag{2.8.2}$$

where the summands υ_{ni}, *defined in* (2.2.4), *are exchangeable binary variables.*

Proof. The order-statistic $U_{(\varphi(i))} \equiv Y_{(\varphi(i))}$ and the concomitant $U_{[i]}$ of $T_{(i)} \equiv X_{(i)}$ can be written in terms of the ranks of T's and U's as follows:

$$U_{(\varphi(i))} = \sum_{\alpha=1}^{n} U_\alpha I_{[S_\alpha = \varphi(i)]}, \qquad (2.8.3)$$

$$U_{[i]} = \sum_{\alpha=1}^{n} U_\alpha I_{[R_\alpha = i]} \, . \qquad (2.8.4)$$

Note that $N(\varphi,\varepsilon)$ is simply a count of how many pairs in the merged file based on φ, namely

$$[\{T_{(i)}, U_{(\varphi(i))}\}, i = 1, 2,\dots, n] \qquad (2.8.5)$$

satisfy the inequality

$$| U_{(\varphi(i))} - U_{[i]} | \leq \varepsilon \, . \qquad (2.8.6)$$

From equation (2.8.4), it is clear that $U_{[i]} = U_j$ iff $R_j = i$. Hence if (2.8.6) holds for some i, then $\exists$ a j such that

$$| U(\varphi(R_j)) - U_j | \leq \varepsilon \, .$$

In view of the continuity of (T_i,U_i), this correspondence is one-to-one. Therefore, the count $N(\varphi,\varepsilon)$ is same as the count given by

$$\sum_{\alpha=1}^{n} I_{[\,| U(\varphi(R_\alpha)) - U_\alpha | \leq \varepsilon\,]} \qquad (2.8.7)$$

Hence, (2.8.2) follows from (2.8.7) and the definition (2.2.4) of $\upsilon_{ni}(\varphi,\varepsilon)$.

In order to show the exchangeability of the summands υ_{ni} in (2.8.2), we note that the original unobserved pairs $W_1,W_2,\dots,W_n$ are independent and identically distributed vectors. Therefore

$$\{ W_{\alpha_1}, W_{\alpha_2},\dots,W_{\alpha_n} \} \stackrel{d}{=} \{ W_1,W_2,\dots,W_n \} \qquad (2.8.8a)$$

where $(\alpha_1, \alpha_2, \dots \alpha_n)$ is an arbitrary permutation of $(1, 2,\dots, n)$.

Define a function $\mathbf{f} \equiv (f_1, f_2 ,\dots, f_n)$ from $\mathbb{R}^{2n}$ to $\mathbb{R}^n$ by the equations

$$f_j(a_1, b_1, \ldots, a_n, b_n) = \begin{cases} 1 & \text{if } \zeta_j(\mathbf{b}, \varepsilon) \le \varphi(\zeta_j(\mathbf{a}, 0)) \le \zeta_j(\mathbf{b}, -\varepsilon) \\ 0 & \text{otherwise} \end{cases} \tag{2.8.8b}$$

for $j = 1, 2, \ldots, n$, where $(a_1, b_1, \ldots, a_n, b_n)$ is an arbitrary point in $\mathbb{R}^{2n}$, $\mathbf{a} = (a_1, \ldots, a_n)$, $\mathbf{b} = (b_1, \ldots, b_n)$, φ is the matching strategy we started with, and

$$\zeta_j(\mathbf{a}, \varepsilon) = \sum_{i=1}^{n} I_{[a_j - a_i \ge \varepsilon]}, \ \forall\, \varepsilon \ge 0. \tag{2.8.8c}$$

It follows from (2.8.8a) and Theorem 2.6.1 that

$$\mathbf{f}(W_{\alpha_1}, W_{\alpha_2}, \ldots, W_{a_n}) \stackrel{d}{=} \mathbf{f}(W_1, W_2, \ldots, W_n). \tag{2.8.9}$$

Fix $j \in \{1, 2, \ldots, n\}$. Then, it is clear that $f_j(W_{\alpha_1}, W_{\alpha_2}, \ldots, W_{\alpha_n})$ is the indicator function of the event

$$[\, \zeta_j(U_\alpha, \varepsilon) \le \varphi(\zeta_j(T_\alpha, 0)) \le \zeta_j(U_\alpha, -\varepsilon)] \,.$$

or, equivalently, in terms of the ranks $R_1, \ldots, R_n$ of the T's and the empirical CDF $G_n(\cdot)$ of U's, the indicator function of the event

$$[G_n(U_{\alpha_j} - \varepsilon) \le (\varphi(R_{\alpha_j}))/n \le G_n(U_{\alpha_j} + \varepsilon)].$$

Since $G_n^{-1}(k/n) = U_{(k)}$, $k = 1, 2, \ldots, n$, it follows that $f_j(W_{\alpha_1}, W_{\alpha_2}, \ldots, W_{\alpha_n})$ is 1 iff $| U_{\varphi((R_{\alpha_j}))} - U_{\alpha_j} | \le \varepsilon$. Consequently,

$$f_j(W_{\alpha_1}, W_{\alpha_2}, \ldots, W_{\alpha_n}) = \upsilon_{n\alpha_j}(\varphi, \varepsilon). \tag{2.8.10a}$$

Similarly,

$$f_j(W_1, \ldots, W_n) = \upsilon_{nj}(\varphi, \varepsilon). \tag{2.8.10b}$$

The exchangeability of the summands in (2.8.2) follows from (2.8.9), (2.8.10a) and (2.8.10b).

We shall now fix the parameters describing dependence in the population of W, and allow n to tend to infinity in order to study the behavior of $N(\varphi^*, \varepsilon)$. In view of the fact that federal files often consist of a large number of records, it is clear that these asymptotic investigations are useful. Later, in this section, we shall also present the results of a Monte Carlo study for $N(\varphi^*,\varepsilon)$, in which we vary the dependence parameters for various fixed values of n. For brevity, we let $N(\varepsilon) = N(\varphi^*,\varepsilon)$. We start with a review of Yahav (1982)'s results concerning $E[N(\varepsilon)] = \mu_n(\varepsilon)$.

Assuming that the distribution of T and U is such that the conditional distribution of U given that T=t is (univariate) normal with mean t and variance 1, Yahav derived the limiting value of $\mu_n(\varepsilon) = E[N(\varepsilon)/n]$ as $n \to \infty$ by using the representation (2.8.1) in which the summands are functions of the order-statistics of $U_1,\ldots, U_n$ and the concomitants of the order-statistics of $T_1,\ldots, T_n$. His proof relied on an approximation theorem for the order-statistics, given in Bickel and Yahav(1977), of observations based on the above model. Furthermore, he also reported the findings of a Monte-Carlo study for $\mu_n(\varepsilon)$ in a particular case of his model, namely, T and U are bivariate normal with correlation ρ.

We first establish the large-sample behavior of $N(\varepsilon)/n$ in case of samples from an *arbitrary population.* The properties of its expected value follow as a consequence. Second, we indicate how Yahav's simulation study of the small-sample properties of $\mu_n(\varepsilon)$ can be improved upon. We shall then present the results of our Monte-Carlo study of $\mu_n(\varepsilon)$ when n takes fixed values.

Theorem 2.8.2. *For an absolutely continuous distribution of* (T,U), *let*

$$\mu(\varepsilon) = P[\, G(U-\varepsilon) \le F(T) \le G(U+\varepsilon)\,]. \tag{2.8.11}$$

Then,

$$N(\varepsilon)/n \to_{Pr} \mu(\varepsilon), \quad \text{as } n \to \infty. \tag{2.8.12}$$

Proof. Let $L_n = N(\varepsilon)/n$. Using the definitions of $A_{ni}(\varepsilon)$ in (2.2.2), and recalling the representation (2.8.2) for $N(\varepsilon)$ as a sum of exchangeable variables, we obtain

$$N(\varepsilon) = \sum_{i=1}^{n} I_{[A_{ni}(\varepsilon)]}.$$

It follows that

$$E(L_n) = n\,P(A_{n1}(\varepsilon))/n = P(A_{n1}(\varepsilon)), \qquad (2.8.13)$$

and

$$E(L_n^2) = n^{-2}\,[\,E(N(\varepsilon)^{(2)}) + E(N(\varepsilon))\,] \qquad (2.8.14)$$

where $E(N(\varepsilon)^{(2)})$ is the second factorial moment of $N(\varepsilon)$. Using the exchangeable representation (2.8.2) again, we get

$$E(L_n^2) = n^{-2}\,[n^{(2)}\,P\{A_{n1}(\varepsilon)\,A_{n2}(\varepsilon)\} + n\,P(A_{n1}(\varepsilon))]. \qquad (2.8.15)$$

For $\alpha = 1,2,\ldots,n$, let

$$\eta_{1\alpha} = \zeta_\alpha(\mathbf{U},\varepsilon) - \zeta_\alpha(\mathbf{T},0)$$

and

$$\eta_{2\alpha} = \zeta_\alpha(\mathbf{T},0) - \zeta_\alpha(\mathbf{U},\varepsilon) \qquad (2.8.16)$$

where $\zeta_\alpha(\cdot,\cdot)$ are defined in (2.8.8c). Using (2.8.16), we get

$$A_{n1}(\varepsilon) = (\eta_{11}/n \le 0, \eta_{21}/n \le 0), \qquad (2.8.17)$$

and

$$A_{n1}(\varepsilon)\,A_{n2}(\varepsilon) = \bigcap_{i=1}^{2}\bigcap_{j=1}^{2} (\eta_{ij}/n \le 0). \qquad (2.8.18)$$

Note that, given $\mathbf{W}_1 = (t_1,u_1)$, the infinite sequence $\xi_{11i} = I_{[U_1 - U_i \ge \varepsilon]} - I_{[T_1 - T_i \ge 0]}$, $i = 2,\ldots,$ is exchangeable. Hence, by the Strong Law of Large Numbers for exchangeable random variables [see Chung, 1974, Chow and Teicher,1978, p.223],

$$\eta_{11}/n \xrightarrow{a.s.} E(\xi_{112} \mid \mathbf{W}_1) \quad \text{as } n \to \infty, \qquad (2.8.19)$$

The conditional expectation on r.h.s of (2.8.19) is equal to $\{G(u_1 - \varepsilon) - F(t_1)\}$. It follows that

$$\eta_{11}/n \xrightarrow{a.s.} G(U_1 - \varepsilon) - F(T_1), \quad \text{as } n \to \infty. \qquad (2.8.20)$$

We can show by similar arguments that

$$\eta_{1\alpha}/n \xrightarrow{a.s.} G(U_\alpha - \varepsilon) - F(T_\alpha), \qquad (2.8.21)$$

and

$$\eta_{2\alpha}/n \xrightarrow{a.s.} F(T_\alpha) - G(U_\alpha + \varepsilon), \qquad (2.8.22)$$

where $\alpha = 1,2$. Using the fact [see Serfling, 1980, p.52] that a sequence of vectors converges almost surely to a given vector iff the component wise sequences converge almost surely to the appropriate components of the limit, we get from (2.8.21) and (2.8.22) that

$$\begin{bmatrix} \eta_{11}/n \\ \eta_{21}/n \\ \eta_{12}/n \\ \eta_{22}/n \end{bmatrix} \overset{a.s}{\rightarrow} \begin{bmatrix} G(U_1-\varepsilon) - F(T_1) \\ F(T_1) - G(U_1+\varepsilon) \\ G(U_2-\varepsilon) - F(T_2) \\ F(T_2) - G(U_2+\varepsilon) \end{bmatrix}. \qquad (2.8.23)$$

It follows from (2.8.17), (2.8.18), (2.8.23) and the independence of W_1 and W_2 that

$$P(A_{n1}(\varepsilon)) \rightarrow \mu(\varepsilon), \quad (2.8.24)$$

and

$$P(A_{n1}(\varepsilon)\, A_{n2}(\varepsilon)) \rightarrow \mu^2(\varepsilon). \qquad (2.8.25)$$

Using (2.8.13), (2.8.14), (2.8.24) and (2.8.25), it is easy to verify that as $n \rightarrow \infty$,

$$E(L_n) \rightarrow \mu(\varepsilon), \text{ and } Var(L_n) \rightarrow 0. \qquad (2.8.26)$$

It is well known that (2.8.26) implies the convergence in probability as in (2.8.12).

The following corollary generalizes Yahav's result concerning $\mu_n(\varepsilon)$, the first moment of $N(\varepsilon)/n$.

Corollary 2.8.1. *For* $p > 0$,

(i) $$N(\varepsilon)/n \xrightarrow{Lp} \mu(\varepsilon) \quad \text{as } n \rightarrow \infty \qquad (2.8.27)$$

(ii) $$E[(N(\varepsilon)/n)^p] \rightarrow [\mu(\varepsilon)]^p \quad \text{as } n \rightarrow \infty. \qquad (2.8.28)$$

Proof. The number of ε-correct matches can at most be n, the number of pairs in the unobserved bivariate-data. Hence, $0 \le N(\varepsilon)/n \le 1$, for all n = 2,3,... and $\{N(\varepsilon)/n\}$ is a uniformly bounded sequence of random variables. It is well known that convergence in probability and L_p-convergence are equivalent for such sequences. Hence (i) follows easily from Theorem 2.8.2. Now, (ii) readily follows from (i) because $(\mu(\varepsilon)^p)$ is finite for p>0.

Note that in Theorem 2.8.2 or Corollary 2.8.1, no assumption about the conditional distribution of U given T has been made as was the case with Yahav's results. Yahav used simulated samples from a bivariate - normal population with mean vector 0 and covariance matrix

$$\Sigma = (1-\rho^2)^{-1} \begin{bmatrix} \rho^2 & \rho^2 \\ \rho^2 & 1 \end{bmatrix}, \qquad (2.8.29)$$

to study small sample properties of $\mu_n(\varepsilon)$. It is important to note that in (2.8.29), the variances of T and U are functions of their correlation, ρ. This is so, because Yahav's model requires that the conditional distribution of U given T=t be normal with mean t and variance 1. The limiting value of $\mu_n(\varepsilon)$ for his particular model is given by:

$$\mu(\varepsilon) = \int_{\mathbb{R}} \{\Phi(x\theta + \varepsilon/\rho) - \Phi(x\theta - \varepsilon/\rho)\}\, d\Phi(x), \qquad (2.8.30)$$

where $\theta = [(1-\rho)/(1+\rho)]^{1/2}$, and Φ denotes the cdf of a standard normal random variable.

Yahav computed $\mu(\varepsilon)$ by numerical integration for ε = 0.01, 0.05, 0.1 & 0.3. However, it can be shown that (2.8.30) simplifies to

$$\mu(\varepsilon) = 1 - 2\,\Phi\,[-((1+\rho)/2)^{-1/2})\,\varepsilon/\rho\,]. \qquad (2.8.31)$$

Yahav also provided Monte-Carlo estimates of $\mu_n(\varepsilon)$, for n = 10, 20,50 using the simulated data on T and U. Table 2.1, given in Appendix A, is a typical example of one of his results.

It is clear from Table 2.1 and equation (2.8.31) that $\mu_n(\varepsilon)$ and $\mu(\varepsilon)$ *decrease* as ρ ranges from 0.001 to 0.99. In fact , (2.8.31) implies that $\mu(\varepsilon)=1-2\Phi(-\varepsilon)$ for $\rho=1.0$ and $\mu(\varepsilon) = 1.0$ for $\rho = 0$. These conclusions go against the intuition, since one expects that for an optimal strategy, such as φ^*, $\mu_n(\varepsilon)$ as well as $\mu(\varepsilon)$ must be monotone increasing in ρ . The problem here is not with the MLP φ^*, but with the covariance matrix Σ, defined by (2.8.29), used in Yahav's model. Because, as ρ changes its value, so do the marginal variances of T and U. In fact, as $\rho \to 1$, the marginal variances $\to \infty$. To rectify this problem, we assume a bivariate normal model for T and U with means zero, variances one and the correlation ρ.

For each combination of four values of n, namely 10, 20, 50 and 100, and twelve values of ρ, namely 0.00, 0.10 (0.10) 0.90, 0.95, 0.99, 1000 independent samples were generated from the bivariate normal population using the IMSL Library routines. These data were used to obtain Monte-Carlo estimates of $\mu_n(\varepsilon)$, where ε was given the values 0.01, 0.05, 0.1, 0.3, 0.5, 0.75, 1.0. Furthermore, it is easy to show that, for the above model

$$\mu(\varepsilon) = P(|Z| \leq \varepsilon / (2(1-\rho))^{1/2}), \qquad (2.8.32)$$

where Z is a standard normal random variable. It is clear from (2.8.32) that $\mu(\varepsilon)$ is a monotone increasing function of ρ . Using standard-normal CDF tables, $\mu(\varepsilon)$ in (2.8.32) was computed for each combination of the twelve values of ρ and the seven values of ε mentioned above. The estimated values of $\mu_n(\varepsilon)$ and the limiting value $\mu(\varepsilon)$ are presented in Table 2.2 - Table 2.8 in Appendix A.

Note that, as expected, $\mu_n(\varepsilon)$ in Tables 2.2 - 2.8 are monotone increasing functions of ρ for each fixed ε. Furthermore, the quality of the merged file is quite good if we want to reconstruct contingency tables with intervals of size .5 σ or more and the correlation $\rho \geq 0.5$.

2.9. Poisson Convergence of N(φ^*)

For the card-matching problem, some distributional properties of the number of correct matches in randomly arranging one pack of cards against another were stated in Proposition 2.3.1. In particular, the well-known approximation of the distribution of the number of correct matches by a Poisson distribution with mean 1 was mentioned. This approximation is motivated by the observation that the occurrence of a match tends to be a rare event when the number of cards in the matching problem grows indefinitely. Inspired by this result, it is natural to ask whether Poisson distributions can approximate the distribution of the number of correct matches in model-based matching strategies. The answer is in the affirmative for the maximum likelihood pairing φ^*. In this section, we examine the Poisson convergence of $N(\varphi^*)$.

The number of correct matches $N(\varphi^*) \equiv N$ is given by

$$N = \sum_{i=1}^{n} I_{A_{ni}} \tag{2.9.1}$$

where $A_{ni} = [R_i = S_i]$, $i = 1, 2, \ldots, n$ are exchangeable events. It follows that $E(N) = nP(A_{n1})$. Zolutikhina and Latishev (1978) *sketched* a proof of the fact that the expectation of N converges to a constant as n tends to ∞. Their approach is to first write $P(A_n)$ as the triple integral

$$\frac{1}{\pi} \int_{-\infty}^{\infty} \int_{-\infty}^{\infty} \int_{\theta=0}^{\pi} \exp\{(n-1)\ln(s(x,y,\theta))\}\, d\theta\, dH(x,y),$$

where $s(x,y,\theta) = p_3(x,y) + 2\sqrt{p_1(x,y)p_2(x,y)} \cdot \cos 2\theta$, $p_1(x,y) = F(x) - H(x,y)$,

$p_2(x,y) = G(y) - H(x,y)$, and $p_3(x,y) = 1 - p_1(x,y) - p_2(x,y)$, $\forall\, x,y \in R$.

Using the Laplace expansion (Bleistein and Handlesman, 1975) of this integral in powers of n^{-1}, they concluded that $P(A_{n1}) \approx \lambda/n$ for large n, the constant λ being

$$\lambda = \int_{-\infty}^{\infty} [h(x, G^{-1} F(x))/h_2(G^{-1}(F(x))]\, dx \tag{2.9.2}$$

where $h_2(\cdot)$ denotes the marginal density of Y. Thus, in large samples, $E(N) \approx \lambda$.

In this section, we shall generalize the result of Zolutikhina and Latishev (1978) by showing that the dth factorial moment of N, $E(N^{(d)})$ converges to λ^d for $d \geq 1$ under certain conditions on the distribution of **W**. The weak convergence of N to the Poisson distribution with mean λ follows as a consequence.

Let us first note that the ranks $\mathbf{R} = (R_1,..., R_n)$ and $\mathbf{S} = (S_1,..., S_n)$ are invariant under monotone functions of T and U respectively. Therefore N is also invariant under such transformations. Without loss of generality, we can replace T and U by F(T) and G(U) respectively. This probability integral transformation allows us to assume that T and U are marginally uniform random variables and that the joint CDF of T and U is a copula $C(\cdot,\cdot)$ (Schweizer and Wolff (1981)). It is well known that the following Frechét bounds apply to any copula:

$$\max(x+y-1,0) \leq C(x,y) \leq \min(x,y), \quad \forall (x,y) \in [0,1]^2 \tag{2.9.3}$$

However, the distribution of N will be derived only for a part of the spectrum of all possible copulas given by (2.9.3). The extreme copula $C(x,y) = \min(x,y)$ is not considered because, for this copula, $N = n$ almost surely. Furthermore, Goel (1975) has observed that $\varphi^* = (1, 2,..., n)$ is the M.L.P. for all data sets iff the joint density of T and U has the M.L.R. property. The M.L.R. property necessarily implies that the corresponding copula must satisfy the inequality $C(x,y) \geq xy$, for all (x,y) in the unit square (Tong (1980), p. 80). We shall henceforth assume that the joint CDF of T and U satisfies the inequalities

$$xy \leq C(x,y) < \min(x,y), \quad \forall (x,y) \in [0,1]^2 \tag{2.9.4}$$

Note that all C(x,y) in (2.9.4), have positive quadrant dependence property and T and U are independent iff $C(x,y) \equiv xy$. We will denote the joint density function that corresponds to the copula C(x,y) by c(x,y).

It is easy to show that the integral in (2.9.2) can now be expressed as

$$\lambda = \int_0^1 c(x,x)\, dx. \tag{2.9.5}$$

We will now prove that N converges weakly to a Poisson distribution with mean λ. First, we state the equivalence between the card matching problem and the M.L.P. in the independence case.

Proposition 2.9.1. *Let* T *and* U *be independent random variables. Then the distribution of* $\underset{\sim}{\nu} = (\upsilon_1, \ldots, \upsilon_n)$ *defined in* (2.2.6) *is the same as that of the vector* $\underset{\sim}{\delta} \equiv (\delta_1, \ldots, \delta_n)$ *where*

$$\delta_i = I_{[R_i = i]}, \quad i = 1, 2, \ldots, n . \tag{2.9.6}$$

It readily follows from Proposition 2.9.1 that, if T and U are independent random variables, then

$$\sum_{i=1}^{n} \upsilon_i \overset{d}{=} \sum_{i=1}^{n} \delta_i = Z_n, \text{ say.} \tag{2.9.7}$$

Thus, the exact as well as the asymptotic distribution of N can be derived by studying Z_n, which is same as the number of matches in the card matching problem. As stated in Proposition 2.3.1, the asymptotic distribution of Z_n is Poisson with mean 1. We now present another proof of this well-known result. The novel feature of our proof is that we establish a certain dependence property of $\delta_1, \ldots, \delta_n$ and consequently derive the limiting distribution by using only the first two moments of Z_n rather than all the factorial moments. The outline of the proof is as follows:

(i) Show that δ_i's have a certain positive dependence structure.

(ii) Invoke a theorem due to Newman (1982) to arrive at the Poisson convergence of N in the independence case.

We first review the definitions of some concepts of positive dependence of random variables.

Definition 2.9.1 (Lehmann, 1966). X_1 *and* X_2 *are said to be positive quadrant dependent (PQD) iff*

$$P(X_1 > x_1, X_2 > x_2) \geq P(X_1 > x_1)\, P(X_2 > x_2), \ \forall\, x_1, x_2 \in R. \tag{2.9.8}$$

Definition 2.9.2 (Newman, 1982). $X_1,\ldots, X_n$ *are said to be linearly positive quadrant dependent (LPQD) iff for any disjoint subsets* A,B *of* {1, 2,..., n} *and positive constants* $a_1,\ldots, a_n$,

$$\sum_{k\in A} a_k X_k \quad \text{and} \quad \sum_{k\in B} a_k X_k \quad \text{are PQD.} \tag{2.9.9}$$

Definition 2.9.3 (Esary, Proschan, Walkup, 1967). $X_1,\ldots, X_n$ *are said to be associated iff for every choice of functions* $f_1(X_1,\ldots, X_n)$ *and* $f_2(X_1,\ldots, X_n)$, *which are monotonic increasing in each argument and have finite variance,*

$$\operatorname{cov}(f_1(X_1,\ldots, X_n), f_2(X_1,\ldots, X_n)) \geq 0. \tag{2.9.10}$$

It is well known that association implies the LPQD property. We will now establish that $\delta_1,\ldots, \delta_n$ in (2.9.6) possess a weaker version of the LPQD property.

Lemma 2.9.1. *For* $k = 1, 2,\ldots, n-1$, *let*

$$\eta_k = \sum_{i=1}^{k} \delta_i. \tag{2.9.11}$$

Then η_k and δ_n are PQD

Proof. Fix $k \in \{1, 2,\ldots, n-1\}$. Then, by Definition 2.9.1, η_k and δ_n are PQD iff

$$P(\eta_k > x_1, \delta_n > x_2) \geq P(\eta_k > x_1)\, P(\delta_n > x_2),\ \forall\, x_1, x_2 \in R_+ \tag{2.9.12}$$

Because δ_i's are binary random variables, we have

$$P(\delta_n > x) = \begin{cases} 1 & \text{if } x < 0 \\ 0 & \text{if } x \geq 1. \end{cases} \tag{2.9.13}$$

It is clear from (2.9.13) that (2.9.12) holds for any $x_2 \notin [0,1)$. For $0 \leq x_2 < 1$, $(\delta_n > x_2) \equiv (\delta_n = 1)$. It therefore remains to be shown that, $\forall\, \ell = 0, 1,\ldots, k$

$$P(\eta_k \geq \ell\,,\, \delta_n = 1) \geq P(\eta_k \geq \ell\,)\ \ P(\delta_n = 1). \tag{2.9.14}$$

Using the fact that

$$P(\eta_k \geq \ell\,) = P(\eta_k \geq \ell\,,\, \delta_n = 0) + P(\eta_k \geq \ell\,,\, \delta_n = 1) \tag{2.9.15}$$

and that $P(\delta_n = 1) = 1/n$, we can rewrite (2.9.14) in a more useful form

$$P(\eta_k \geq \ell \,|\, \delta_n = 0) \leq P(\eta_k \geq \ell \,|\, \delta_n = 1),\ \ell = 0,\ 1,2,\ldots,k. \tag{2.9.16}$$

We now proceed to establish the inequality in (2.9.16) by means of a combinatorial argument. Since the event $(\delta_n = 0)$ is equivalent to the event $\bigcup_{\alpha=1}^{n-1} (R_n{=}\alpha)$, we can write

$$(\eta_k \geq \ell\,,\, \delta_n = 0) = \bigcup_{\alpha=1}^{n-1} J_\alpha, \tag{2.9.17}$$

where

$$J_\alpha = (\eta_k \geq \ell\,,\, R_n{=}\alpha),\ \alpha = 1,\ 2,\ldots,\ (n\text{-}1). \tag{2.9.18}$$

Note that J_α's in (2.9.18) are mutually disjoint measurable subsets of Φ. Let us now fix α as well. Then, any permutation φ in J_α satisfies $\varphi(n) = \alpha$ and $(\varphi(1),\ldots, \varphi(n\text{-}1))$ is an arrangement of the integers $1, 2,\ldots, \alpha\text{-}1, \alpha\text{+}1,\ldots, n$ producing at least ℓ matches of the type $\varphi(i) = i$ in the positions $i = 1, 2,\ldots, k$. On the other hand, any permutation φ in $(\eta_k \geq \ell, \delta_n = 1)$ satisfies $\varphi(n) = n$ and $(\varphi(1),\ldots, \varphi(n\text{-}1))$ is an arrangement of the integers $1, 2,\ldots, n\text{-}1$ yielding at least ℓ matches such as $\varphi(i) = i$ in the positions $i = 1, 2,\ldots, k$. Because $\alpha \neq n$, it can be shown that

$$\#(J_\alpha) \leq \#(\eta_k \geq \ell,\, \delta_n = 1), \tag{2.9.19}$$

where $\#(A)$ denotes the cardinality of the set A. Since α, k and ℓ are arbitrary, (2.9.17) implies that for $k = 1, 2,\ldots, n\text{-}1;\ \ \ell = 0,\ldots, k$,

$$\#(\eta_k \geq \ell,\, \delta_n = 0) \leq (n\text{-}1)\ \#(\eta_k \geq \ell,\, \delta_n = 1)\,. \tag{2.9.20}$$

Since the random variable **R** is discrete uniform on Φ, it follows from (2.9.20) that

$$P(\eta_k \geq \ell\,,\, \delta_n = 0) \leq (n-1)\, P(\eta_k \geq \ell\,,\, \delta_n = 1)\,. \qquad (2.9.21)$$

Multiplying both sides of the inequality in (2.9.21) by n and using (2.9.15) we obtain (2.9.16).

For our purposes here, the weaker version of LPQD established in Lemma 2.9.1 is sufficient. In Goel and Ramalingam (1985), we had also conjectured that $\delta_1, \ldots, \delta_n$ are, in fact, associated random variables. This conjecture has recently been established by Fishburn et.al.(1987). We now state two useful results due to Newman (1982). Let $\psi_{X_1,\ldots,X_n}(r_1,\ldots,r_n)$ denote the joint characteristic function of $X_1,\ldots, X_n$.

Lemma 2.9.2. *If* X_1 *and* X_2 *are PQD, then*

$$|\psi_{X_1,X_2}(r_1,r_2) - \psi_{X_1}(r_1)\,\psi_{X_2}(r_2)| \leq |r_1 r_2|\,\mathrm{cov}(X_1,X_2),\ \forall\, r_1,r_2 \in R \quad (2.9.22)$$

Lemma 2.9.3. *Suppose that* $X_1,X_2,\ldots, X_n$ *are LPQD. Then*

$$|\psi_{X_1,\ldots,X_n}(r_1,\ldots,r_n) - \prod_{j=1}^{n} \psi_{X_j}(r_j)| \leq \sum_{k=1}^{n} \sum_{\substack{j=1 \\ k<j}}^{n} |r_k r_j|\,\mathrm{cov}(X_k,X_j),$$

$$\forall\, r_1,\ldots,r_n \in R. \qquad (2.9.23)$$

If $X_1,\ldots,X_n$ are exchangeable random variables, and $r_1 = r_2 = \ldots = r_n = r$, then it follows from (2.9.23) that

$$|\psi_{\Sigma X_i}(r) - \psi^{n}_{X_1}(r)| \leq \frac{n(n-1)}{2}|r|^2\,\mathrm{cov}(X_1, X_2) \qquad (2.9.24)$$

In general, this estimate of error in approximating the characteristic function of $\sum_{i=1}^{n} X_i$ by the product of the marginal characteristic functions is valid when $X_1,\ldots, X_n$ are LPQD. However, with regard to the binary variables $\delta_1,\ldots, \delta_n$, an estimate similar to (2.9.24) can be obtained under a weaker version of the LPQD property mentioned in Lemma 2.9.1.

Lemma 2.9.4. *Let* δ_i*'s be the Bernoulli variables in* (2.9.6) *and let* Z_n *be as defined in* (2.9.7). *Then,*

$$|\Psi_{Z_n}(r) - \Psi^n_{\delta_1}(r)| \le \frac{n(n-1)}{2}|r|^2 \operatorname{cov}(\delta_1,\delta_2),\ \forall\, n \ge 2,\ r \in R, \qquad (2.9.25)$$

Proof. Since $\delta_1,\ldots,\delta_n$, are exchangeable variables we have

$$\operatorname{cov}(\delta_i,\delta_j) = \operatorname{cov}(\delta_1,\delta_2),\ \forall\, i \ne j \qquad (2.9.26)$$

$$\Psi_{\delta_j}(r) = \Psi_{\delta_1}(r),\ \forall j. \qquad (2.9.27)$$

Fix $n \ge 2$ and consider the following finite sequence of statements:

$$|\Psi_{\eta_k}(r) - \Psi^k_{\delta_1}(r)| \le \frac{k(k-1)}{2}|r|^2 \operatorname{cov}(\delta_1,\delta_2),\ \text{for } k = 2,3,\ldots,n \qquad (2.9.28)$$

Note that (2.9.25) is obtained from (2.9.28) by letting $k=n$. We shall now establish (2.9.28) by induction on k.

Since δ_1 and δ_n are PQD, Lemma 2.9.2 readily implies that (2.9.28) holds for $k = 2$. Now, let us assume that (2.9.28) holds for some $k = m - 1$ where $3 < m < n$. In order to establish (2.9.28) for $k=m$, we note that $\eta_m = \eta_{m-1} + \delta_m$ and that, in view of the exchangeability of $(\delta_1,\ldots,\delta_n)$ and Lemma 2.9.1, η_{m-1} and δ_m are PQD random variables. Hence using Lemma 2.9.2, we obtain

$$|\Psi_{\eta_m}(r) - \Psi_{\eta_{m-1}}(r)\Psi_{\delta_m}(r)| \le |r|^2 \operatorname{cov}(\eta_{m-1},\delta_m), \qquad (2.9.29)$$

It follows from (2.9.26) that the right hand side in the above inequality can be written as

$$|r|^2 (m-1) \operatorname{cov}(\delta_1,\delta_2). \qquad (2.9.30)$$

To complete the proof by induction, note that

$$|\Psi_{\eta_m}(r)-\Psi^m_{\delta_1}(r)| \le |\Psi_{\eta_m}(r)-\Psi_{\eta_{m-1}}(r)\Psi_{\delta_m}(r)| + |\Psi_{\eta_{m-1}}(r)\Psi_{\delta_m}(r)-\Psi^m_{\delta_1}(r)|.$$

Using (2.9.29) and (2.9.30) and the fact that $|\Psi_{\delta_1}(r)| \le 1$, the right hand side in the above expression is $\le |r|^2 (m-1) \operatorname{cov}(\delta_1,\delta_2) + |\Psi_{\eta_{m-1}}(r) - \Psi^{m-1}_{\delta_1}(r)|$. By the induction hypothesis (2.9.28) holds for $k=m-1$. Therefore, the bound in the last inequality becomes

$$|r|^2(m-1)\mathrm{cov}(\delta_1,\delta_2)+|r|^2\frac{(m-1)(m-2)}{2}\mathrm{cov}(\delta_1,\delta_2)=\frac{m(m-1)}{2}|r|^2\mathrm{cov}(\delta_1,\delta_2). \quad (2.9.31)$$

The proof of (2.9.28) is now complete by the induction argument.

The Poisson convergence of N in the independence case, can now be established as follows:

Theorem 2.9.1. *Let* T *and* U *be independent random variables. Let the number of correct matches,* N, *be given by* (2.9.1). *Then*

$$N \xrightarrow{d} \text{Poisson }(1), \text{ as } n \to \infty \quad (2.9.32)$$

Proof: By equation (2.9.7), $N \stackrel{d}{=} Z_n$. Furthermore,

$$\mathrm{cov}(\delta_1,\delta_2) = P(R_1 = 1, R_2 = 2) - [P(R_1 = 1)]^2. \quad (2.9.33)$$

Since $P(R_1 = 1, R_2 = 2) = 1/n(n-1)$, it follows that

$$n(n-1)\,\mathrm{cov}(\delta_1,\delta_2) = \frac{1}{n},\ \forall\, n \geq 2, \quad (2.9.34)$$

The proof of (2.9.32) requires that the characteristic function of Z_n converge to the characteristic function of the Poisson distribution with mean 1, i.e.,

$$\beta_n(r) = |\Psi_{Z_n}(r) - \exp(e^{ir} - 1)| \to 0 \text{ as } n \to \infty,\ \forall\, r \in R \quad (2.9.35)$$

Now,

$$\beta_n(r) \leq |\Psi_{Z_n}(r) - \Psi^n_{\delta_1}(r)| + |\Psi^n_{\delta_1}(r) - \exp(e^{ir} - 1)|$$

From Lemma 2.9.4, we obtain

$$\beta_n(r) \leq \frac{n(n-1)}{2}|r|^2\,\mathrm{cov}(\delta_1,\delta_2) + |\Psi^n_{\delta_1}(r) - \exp(e^{ir} - 1)| \quad (2.9.36)$$

Now, the characteristic function of δ_1 is given by

$$\Psi_{\delta_1}(r) = [1 + \frac{1}{n}(\exp(ir) - 1)].$$

Clearly,

$$\Psi^n_{\delta_1}(r) \to \exp(\exp(ir) - 1),\ \forall\, r \in R, \text{ as } n \to \infty \quad (2.9.37)$$

It readily follows from (2.9.34), (2.9.36) and (2.9.37) that (2.9.35) holds.

We now assume that the broken random sample comes from a population in which T and U are dependent random variables. It should be noted that extensions of some of the techniques used in the proof of the Poisson convergence in the independence case to the dependence case are not available at this time. Specifically, no proof of the counterpart of Lemma 2.9.1, that is.

$$\sum_{i=1}^{k} \upsilon_i \text{ and } \upsilon_n \text{ are PQD } \forall\ k=1, 2, \ldots, (n-1) \qquad (2.9.38)$$

is known. Direct verification of the association of $\upsilon_1, \ldots, \upsilon_n$ has been carried out for n = 2, 3, 4 when T and U have the Morgenstern distribution given by (2.6.6). In view of the proof of the association of $\upsilon_1, \ldots, \upsilon_n$ in the independence case due to Fishburn et.al (1987), it is natural to conjecture that the property (2.9.38) of $\upsilon_1, \ldots, \upsilon_n$ holds under certain positive dependence structures for random variables T and U.

In the absence of a valid proof of Lemma 2.9.1 in the dependent case, we need extra conditions on the distribution of T and U in order to derive the Poisson convergence of N. The following lemma will be useful in deriving the main result of this section.

Lemma 2.9.5. *For a fixed d, let* $\mathbf{L}_n = \frac{\mathbf{B}_n}{n}$ *and* $\mathbf{L} = (L_1, \ldots, L_d)'$, *where* $\mathbf{B}_n$ *is defined in (2.2.10) and* $L_j = T_j - U_j$, $j = 1, \ldots, d$. *Then,*

$$\mathbf{L}_n \xrightarrow{a.s.} \mathbf{L}, \text{ as } n \to \infty. \qquad (2.9.39)$$

Proof. Fix $d \geq 1$. It is clear from the definitions of $\underset{\sim}{\xi}_k$ in (2.2.10) and the sigma-field Λ_d in Section 2.2 that the infinite sequence

$$\underset{\sim}{\xi}_{d+1}, \underset{\sim}{\xi}_{d+2}, \ldots, \ldots$$

of d-dimensional vectors are conditionally i.i.d given Λ_d. Hence, using the strong law of large numbers for exchangeable sequences (Chow and Teicher, p. 223) we get

$$\frac{1}{n-d}\sum_{k=d+1}^{n}\underset{\sim}{\xi}_k \overset{a.s.}{\rightarrow} E(\underset{\sim}{\xi}_{d+1} \mid \Lambda_d) \tag{2.9.40}$$

In order to evaluate the limiting conditional expectation in (2.9.40), note first that, for $j = 1,2,...,d$, T_j and U_j are uniform random variables. Now,

$$E(\xi_{jd+1} \mid T_j = t_j,\ U_j = u_j)$$

$$= P(t_j - T_{d+1} \geq 0) - P(u_j - U_{d+1} \geq 0)$$

$$= P(T_{d+1} \leq t_j) - P(U_{d+1} \leq u_j) = t_j - u_j = L_j. \tag{2.9.41}$$

Therefore, it follows from (2.9.41) and the definition of $\underset{\sim}{\xi}_{d+1}$ in (2.2.10) that

$$E(\underset{\sim}{\xi}_{d+1} \mid \Lambda_d) = (L_1, L_2, ..., L_d)'. \tag{2.9.42}$$

Hence, (2.9.40) and (2.9.42) imply that

$$\frac{1}{n-d}\sum_{k=d+1}^{n}\underset{\sim}{\xi}_k \overset{a.s.}{\rightarrow} L, \text{ as } n \rightarrow \infty. \tag{2.9.43}$$

Also, d being a fixed integer, we have

$$\frac{1}{n-d}\sum_{k=1}^{d}\underset{\sim}{\xi}_k \overset{a.s}{\rightarrow} 0, \text{ as } n \rightarrow \infty. \tag{2.9.44}$$

Now, since

$$L_n = \frac{1}{n}\sum_{k=1}^{n}\underset{\sim}{\xi}_k$$

the lemma follows from (2.9.43) and (2.9.44).

The following assumptions will be used to prove the next theorem.

Assumptions. *In the notations of Section* 2.2, *and equation* (2.9.5)

(a) $\lambda < \infty$ (2.9.45)

(b) $\int_{-\infty}^{\infty} |\Psi_L(\theta)|\, d\theta < \infty$ (2.9.46) and (c) $P(\Psi_d^* \leq t) = 0(t^d)$ as $t \to \infty, \forall\, d \geq 1.$ (2.9.47)

Theorem 2.9.2. *If Assumptions* (2.9.45) *to* (2.9.47) *hold, then*

$$N \xrightarrow{d} \text{Poisson }(\lambda) \text{ as } n \to \infty \tag{2.9.48}$$

Proof. Proof of (2.9.48) consists in showing that the factorial moments of N converge to those of the Poisson distribution with mean λ. In other words,

$$E(N^{(d)}) \to \lambda^d, \forall\, d = 1, 2, \ldots . \tag{2.9.49}$$

By the Fourier inversion theorem,

$$P(\mathbf{B}_n = 0) = (2\pi)^{-d} \int_{-\pi}^{\pi} \cdots \int_{-\pi}^{\pi} \Psi_{\mathbf{B}_n}(\underset{\sim}{\theta})\, d\underset{\sim}{\theta}, \tag{2.9.50}$$

where $\psi_{\mathbf{B}_n}(\underset{\sim}{\theta})$ is the characteristic function of the d-dimensional random vector $\mathbf{B}_n$ defined in (2.2.7).

The Assumption (2.9.46) ensures that the Fourier inversion theorem can be applied to the characteristic function of the continuous random variable L. Noting that λ is the value of the density function of L = T - U at 0, we get

$$\lambda = g_L(0) = (2\pi)^{-1} \int_{-\infty}^{\infty} \Psi_L(t)\, dt.$$

Since L_j, j = 1,2,..., d, are i.i.d., with their common density function equal to $g_L(.)$, it follows that

$$\lambda^d = (2\pi)^{-d} \int_{-\infty}^{\infty} \cdots \int_{-\infty}^{\infty} \Psi_L(\underset{\sim}{\theta})\, d\underset{\sim}{\theta} \tag{2.9.51}$$

Recalling that $N = \sum_{i=1}^{n} I_{A_{ni}}$, and $n^{(d)} = n(n-1) \ldots (n-d+1)$,

$$E(N^{(d)}) = n^{(d)} P(A_{n1}A_{n2} \ldots A_{nd})$$

$$= n^{(d)}\ P(B_n = 0). \tag{2.9.52}$$

For fixed d, it is clear that $n^{(d)} \approx n^d$ as $n \to \infty$. It therefore follows from (2.9.52) that, in order to prove (2.9.49), it is sufficient to show that

$$\lim_{n\to\infty} |\Delta(d,n)| = 0, \tag{2.9.53}$$

where $\Delta(d,n) = n^{(d)}\ P(B_n = 0) - \lambda^d$. From (2.9.50) and (2.9.51), we obtain

$$\Delta(d,n) = n^{(d)} (2\pi)^{-d} \int_{-\pi}^{\pi}\cdots\int_{-\pi}^{\pi} \Psi_{S_n}(u)du - (2\pi)^{-d} \int_{-\infty}^{\infty}\cdots\int_{-\infty}^{\infty} \Psi_L(\underset{\sim}{\theta})\, d\underset{\sim}{\theta}. \tag{2.9.54}$$

On making the change of variables $\underset{\sim}{\theta} = (nu_1,\dots, nu_d)$ in the first term of (2.9.54) and noting that $\Psi_{B_n}(\underset{\sim}{\theta}/n) = \Psi_{L_n}(\underset{\sim}{\theta})$, we get

$$\Delta(d,n) = (2\pi)^{-d} \int_{-n\pi}^{n\pi}\cdots\int_{-n\pi}^{n\pi} \Psi_{L_n}(\underset{\sim}{\theta})\, d\underset{\sim}{\theta} - \int_{-\infty}^{\infty}\cdots\int_{-\infty}^{\infty} \Psi_L(\underset{\sim}{\theta})\, d\underset{\sim}{\theta}. \tag{2.9.55}$$

For positive constants α and β, to be determined later, define four integrals as follows:

(i) $$J_1(n) = -\int\cdots\int_{|\underset{\sim}{\theta}|>\alpha} \Psi_L(\underset{\sim}{\theta})\, d\underset{\sim}{\theta} \tag{2.9.56}$$

(ii) $$J_2(n) = \int\cdots\int_{|\underset{\sim}{\theta}|>\alpha} [\Psi_{L_n}(\underset{\sim}{\theta}) - \Psi_L(\underset{\sim}{\theta})]d\underset{\sim}{\theta} \tag{2.9.57}$$

(iii) $$J_3(n) = \int\cdots\int_{\alpha/n\le|\underset{\sim}{\theta}|/n<\beta} \Psi_{L_n}(\underset{\sim}{\theta})d\underset{\sim}{\theta}, \tag{2.9.58}$$

(iv) $$J_4(n) = \int_{\beta n \le |\underset{\sim}{\theta}| \le \pi n} \cdots \int \Psi_{L_n}(\underset{\sim}{\theta})d\underset{\sim}{\theta}. \tag{2.9.59}$$

Using (2.9.54) and the integrals in (2.9.56) to (2.9.59), we can write

$$\Delta(d,n) = (2\pi)^{-d} \sum_{k=1}^{4} J_k. \tag{2.9.60}$$

For appropriate choices of α and β, we will now show that $|J_k(n)| \to 0$ as $n \to \infty$, $k = 1,2,3,4$. Let $\varepsilon > 0$ be a fixed number. Then, Assumption (2.9.45) and the expression (2.9.51) imply that $\Psi_L(\theta)$ is absolutely integrable on R^d. Therefore, we can find a large enough α such that

$$|J_1| \le \int_{|\underset{\sim}{\theta}|>\alpha} \cdots \int |\Psi_L(\underset{\sim}{\theta})| d\underset{\sim}{\theta} < \varepsilon/4 \tag{2.9.61}$$

From Lemma 2.9.5, we have $L_n \xrightarrow{a.s.} L$,which implies that (cf. Bhattacharya and Ranga Rao, 1976, p.44)

$$\Psi_{L_n}(\underset{\sim}{\theta}) \to \Psi_L(\underset{\sim}{\theta}) \quad \text{as } n \to \infty,$$

the convergence being uniform on the compact subset $\{\underset{\sim}{\theta}:\underset{\sim}{\theta} \in R^d \text{ and } |\underset{\sim}{\theta}| \le \alpha\}$. Hence, for the α chosen above, we can find n_1 such that $\forall\, n \ge n_1$,

$$|J_2(n)| < \varepsilon/4. \tag{2.9.62}$$

In order to show that $|J_3(n)| \to 0$, we transform $\underset{\sim}{\theta}$ to $\mathbf{r} = \underset{\sim}{\theta}/n$ in J_3 and obtain

$$J_3(n) = n^d \int_{\alpha/n \le |\mathbf{r}| < \beta} \cdots \int \Psi_{B_n}(\mathbf{r})d\mathbf{r} \tag{2.9.63}$$

Note that $\mathbf{B}_n$ is a lattice random vector so all its moments exist, and that

$$E(\mathbf{B}_n) = \mathbf{0}. \qquad (2.9.64)$$

It was argued in the proof of Lemma 2.9.5 that, for all $n > d$, $\underset{\sim}{\xi}_{d+1},\ldots, \underset{\sim}{\xi}_n$ are conditionally i.i.d. given Λ_d with mean $E(\underset{\sim}{\xi}_j \mid \Lambda_d) = \mathbf{L}$, $\forall\ j = d+1,\ldots, n$. Furthermore, the dispersion matrices $D(\underset{\sim}{\xi}_j \mid \Lambda_d)$, $j = d+1,\ldots, n$, are positive definite. Moreover, for $j = 1,2,\ldots, d$, $\underset{\sim}{\xi}_j$ is degenerate given Λ_d and

$$D(\mathbf{L}) = \sigma^2 \mathbf{I}, \qquad (2.9.65)$$

where $\sigma^2 = \text{var}(L)$ and $\mathbf{I}$ is the dxd identity matrix. The dispersion matrix of $\mathbf{B}_n$ for $n > d$ is,

$$D(\mathbf{B}_n) = E(D(\sum_{i=1}^{n} \underset{\sim}{\xi}_i \mid \Lambda_d)) + D(E(\sum_{i=1}^{n} \underset{\sim}{\xi}_i \mid \Lambda_d))$$

$$= (n-d)\ ED(\underset{\sim}{\xi}_{d+1} \mid \Lambda_d) + (n-d)^2 D(\mathbf{L}).$$

We finally conclude that

$$D(\mathbf{B}_n) - (n-d)^2 \sigma^2 \mathbf{I} = (n-d)\ ED(\underset{\sim}{\xi}_{d+1} \mid \Lambda) \qquad (2.9.66)$$

is positive definite. As the second-order moments of $\mathbf{B}_n$ exist, expand $\Psi_{\mathbf{B}_n}(\mathbf{r})$ around $\mathbf{r} = \mathbf{0}$. Using (2.9.64), we obtain

$$\log \Psi_{\mathbf{B}_n}(\mathbf{r}) = -\frac{1}{2}\mathbf{r}'D(\mathbf{B}_n)\mathbf{r} + 0(\|\mathbf{r}\|^2), \text{ as } \|\mathbf{r}\| \to 0. \qquad (2.9.67)$$

In view of (2.9.65), we obtain

$$|\exp(\log \Psi_{\mathbf{B}_n}(\mathbf{r}))| \le \exp(-\frac{(n-d)^2}{2}\ \sigma^2\|\mathbf{r}\|^2 + 0\|\mathbf{r}\|^2), \text{ as } \|\mathbf{r}\|^2 \to 0.$$

Hence, there exists a constant $\beta > 0$ such that for $n > d$,

$$|\psi_{B_n}(r)| \le \exp(-\frac{1}{4}(n-d)^2 \sigma^2 \|r\|^2),\ \forall \|r\| \le \beta . \qquad (2.9.68)$$

Now, $\exists\ n_2$ such that $\forall\ n \ge n_2, \frac{\alpha}{n} < \beta$ so that, using (2.9.64) and (2.9.68), we obtain

$$|J_3(n)| \le n^d \int_{\alpha/n<|r|<\beta} \cdots \int \exp(-\frac{1}{4}(n-d)^2 \sigma^2 \|r\|^2)\, dr$$

$$\le \int_{|\underset{\sim}{\theta}|>\alpha} \cdots \int \exp(-\frac{1}{4}\sigma^2 \|r\|^2)\, dr . \qquad (2.9.69)$$

It is clear that we can choose a large enough α in (2.9.69) such that $\forall\ n \ge n_2$,

$$|J_3(n)| < \varepsilon/4. \qquad (2.9.70)$$

Finally, to show that $|J_4| \to 0$, we transform $\mathbf{u} = \underset{\sim}{\theta}/n$ in (2.9.59), and obtain

$$|J_4(n)| \le n^d \int_{\beta \le |\mathbf{u}| \le \pi} |\psi_{B_n}(\mathbf{u})|\, d\mathbf{u} . \qquad (2.9.71)$$

In view of the earlier remarks about the conditional distributions of $\underset{\sim}{\xi}_1, \ldots, \underset{\sim}{\xi}_n$ given Λ_d, we obtain for $n \ge d$,

$$|\psi_{B_n}(\mathbf{u})| \le E^{\Lambda_d} |\psi_{\underset{\sim}{\xi}_{d+1}(w_1, \ldots, w_d)}(\mathbf{u})|^{n-d}, \qquad (2.9.72)$$

where $\underset{\sim}{\xi}_{d+1}(\mathbf{w}_1, \ldots, \mathbf{w}_d)$ is the value of $\underset{\sim}{\xi}_{d+1}$ given $\mathbf{W}_i = (T_i, U_i)$, $i = 1, 2, \ldots, d$. Since the characteristic function $\psi_{\underset{\sim}{\xi}_{d+1}}(\mathbf{u})$ is uniformly continuous on the compact set $\{\mathbf{u}: \beta \le |\mathbf{u}| \le \pi\}$ of R^d, it attains its maximum inside this set, say at $\mathbf{u} = \mathbf{u}^*$. Furthermore, $\psi_{\underset{\sim}{\xi}_{d+1}}$ has period 2π so that, for almost all realizations $(\mathbf{w}_1, \ldots, \mathbf{w}_d)$,

$$\sup_{\beta \le |u| \le \pi} |\Psi_{\underset{\sim}{\xi}_{d+1}}(u)| < 1. \tag{2.9.73}$$

Letting $\Psi_d^* = -\ln[\Psi_{\underset{\sim}{\xi}_{d+1}}(u^*)]$, we get from (2.9.71) and (2.9.72),

$$|J_4(n)| \le n^d \, E^{\Lambda_d}[\exp(-(n-d)\Psi_d^*)] = n^d \, M_{\Psi_d^*}(n-d), \tag{2.9.74}$$

where

$$M(s) = \int_0^\infty \cdots \int_0^\infty \exp(-s\Psi_d^*) \prod_{j=1}^{d} dC(x_j,y_j) \tag{2.9.75}$$

is the moment generating function of Ψ_d^* with a real positive argument.

Now, using the Abelian Theorem (cf. Widder (1941), p. 181), we obtain

$$\operatorname*{Lim\,sup}_{t\to\infty} t^d \, M_{\Psi_d^*}(t) \le \operatorname*{Lim\,sup}_{t\downarrow 0} [\, P(\Psi_d^* < t)\, \Gamma(d+1)/t^{\,d}]. \tag{2.9.76}$$

By Assumption (2.9.47), the right-hand side of (2.9.76) is zero and it follows that

$$n^d \, M_{\Psi_d^*}(n-d) \to 0, \text{ as } n \to \infty.$$

In view of (2.9.74), we obtain

$$|J_4(n)| \to 0, \text{ as } n \to \infty. \tag{2.9.77}$$

It follows from (2.9.61), (2.9.62), (2.9.70) and (2.9.77) that $\lim_{n\to\infty} |\Delta(d,n)| = 0$. The convergence of factorial moments in (2.9.49) follows immediately, which in turn implies the Poisson convergence in (2.9.48).

The validity of Theorem 2.9.2 depends on whether the Assumptions (2.9.45) to (2.9.47) hold or not. We shall now give some examples in order to illustrate the fact that these Assumptions are not vacuous. We start with a discussion of the finiteness of λ.

For any copula C(x,y) on $[0,1]^2$, one may define ϕ^2 (possibly an infinite #) by the equation

$$\phi^2 + 1 = \int_0^1 c^2(x,y)\, dx\, dy. \tag{2.9.78}$$

Note that C(x,y) is called a ϕ^2 -bounded distribution if $\phi^2 < \infty$. The class of ϕ^2-bounded distributions is large, as is evident from the following general result.

Proposition 2.9.3 (Lancaster 1969, page 95). *If* H(t,u) *is a* ϕ^2*-bounded distribution with marginal distributions* F(t) *and* G(u) *then complete sets of orthonormal functions* η_{1i}, η_{2i}, i = 1, 2,..., *can be defined on the marginal distributions such that*

$$dH(t,u) = [1 + \sum_{i=1}^{\infty} \rho_i \eta_{1i}(t)\, \eta_{2i}(u)]\, dF(t)\, dG(u) \tag{2.9.79}$$

and

$$\phi^2 = \sum_{i=1}^{\infty} \rho_i^2 . \tag{2.9.80}$$

When T and U are PDE, all ρ_i in the expansion (2.9.79) are non-negative. Therefore, for a ϕ^2-bounded copula C(t,u) with PDE distribution, using the orthonormality of $\{\eta_i\}$, λ in (2.9.5) is given by

$$\lambda = 1 + \sum_{i=1}^{\infty} \rho_i . \tag{2.9.81}$$

For the Morgenstern distribution in (2.6.6), it is known that $\rho_1 = \alpha/3$ with $-1 \le \alpha \le 1$ and $\rho_i = 0$ for $i \ge 2$. Therefore, λ $(= 1 + \alpha/3)$ is finite. Similarly in the bivariate normal distribution with canonical expansion given by (2.6.5), $\lambda = 1/(1-\rho)$,which is finite if $\rho < 1$. Zolutikhina and Latishev (1978) obtained this value of λ via a long argument. Since these important distributions satisfy Assumption (2.9.45), it is obviously not a vacuous assumption.

(Bhattacharya and Ranga Rao, 1976 p. 189-192), provides conditions that are equivalent to the assumption (2.9.46). We cite one here:

Let G_L^{*m} denote the mth convolution of the distribution of L, where $m \geq 1$. If there exists an integer m such that G_L^{*m} has a bounded (almost everywhere) density, then the modulus of the characteristic function of L is integrable on $(-\infty,\infty)$ and vice versa.

Another *sufficient* condition for absolute integrability of $\psi_L(\theta)$ is due to Bochner and Chandrasekar (1949). If there exists a bounded (almost everywhere) density $g_L(t)$ of L and if its characteristic function $\psi_L(\theta)$ is (real) and nonnegative, then

$$\int_{-\infty}^{\infty} |\psi_L(\theta)|\, d\theta < \infty .$$

We illustrate the use of this sufficient (but not a necessary) condition when the PDF of W is given by

$$c(x,y) = 1 + \alpha\,(1 - 2x)(1 - 2y).$$

Clearly, as $|\alpha| \leq 1$, $|x| \leq 1$, $|y| \leq 1$, $\exists$ a positive constant k such that

$$c(x,y) \leq k, \ \forall\ (x,y) \in [0,1]^2 .$$

Note that

$$g_L(t) = \int_0^{1-t} c(t+y,y)\, dy, \quad \forall\ t > 0.$$

By the symmetry of c(x,y) in x and y, it can be shown that $g_L(-t)=g_L(t)$, $\forall\ t > 0$.

Now, using the bound k for c(x,y), and the fact that [-1,1] is the support of L, we get

$$g_L(t) \leq k \int_0^{(1-t)} dy \leq 2k < \infty .$$

Hence, it follows that the PDF of L is (almost everywhere) bounded. We now show that $\psi_L(\theta)$ is real and nonnegative $\forall\ \alpha \geq 0$.

$$\psi_L(\theta) = E(e^{i(T-U)\theta}) = I_1 + \alpha I_2$$

where $$I_1 = \int_0^1 \int_0^1 e^{i(x-y)\theta}\, dxdy = |Z_1|^2$$

and $$I_2 = \int_0^1 \int_0^1 e^{i(x-y)\theta}\,(1-2x)(1-2y)dxdy = |Z_2|^2,$$

with $$Z_1 = \int_0^1 e^{ix\theta}\, dx \text{ and } Z_2 = \int_0^1 e^{ix\theta}\,(1-2x)dx.$$

Hence, $\Psi_L(\theta) = |Z_1(\theta)|^2 + \alpha|Z_2(\theta)|^2 \geq 0$ if $\alpha \geq 0$. Invoking Bochner's sufficient condition, we get $\int_{-\infty}^{\infty} |\Psi_L(\theta)|\, d\theta < \infty$, if $\alpha \geq 0$. However, for all α,

$$\int_{-\infty}^{\infty} |\Psi_L(\theta)|\, d\theta = \int_{-\infty}^{\infty} |Z_1(\theta)|^2\, d\theta + \alpha \int_{-\infty}^{\infty} |Z_2(\theta)|^2, \qquad (2.9.82)$$

so that the two integrals on the right hand side must be finite when $\alpha > 0$. It follows that, even when $\alpha < 0$, $\int_{-\infty}^{\infty} |\psi_L(\theta)|\, d\theta < \infty$. We conclude that (2.9.46) is valid for any member of the Morgenstern family of densities.

Lastly, we discuss the validity of (2.9.47). To be specific, when d=1, one can get the bound

$$|\Psi_{\xi_2(w)}(\theta)| \leq 1 - P_o(1-P_o) + \sin^2(\beta/2)\ \forall\ \beta \leq \theta \leq \pi,\ w = (x,y)'$$

where $P_o = P_o(w) = 1 - x - y + 2C(x,y)$. Therefore,

$$|J_4(n,\beta)| \leq \int_{-\infty}^{\infty}\int_{-\infty}^{\infty} n\, e^{-(n-1)4\sin^2\beta[P_o(1-P_o)]}\, dxdy.$$

Thus, $J_4 \to 0$ as $n \to \infty$ if we show that $nM_{P_o(1-P_o)}(n) \to 0$ as $n \to \infty$, where $M_\eta(s)$ is the Laplace transform of η. A sufficient condition for this to hold is

$$P(P_o(1-P_o) \leq t) = 0(t), \text{ as } t \to 0. \quad (2.9.83)$$

Let $\delta(t)$ and $1-\delta(t)$ be the roots of the equation $P_o(1-P_o) = t$. It suffices to show, as $t \to 0$,

$$P(P_o \leq \delta(t)) = 0(t) \quad (2.9.84)$$

and

$$P(P_o \geq 1 - \delta(t)) = 0(t). \quad (2.9.85)$$

If T and U are independent random variables, then the PDF of P_o can be shown to be

$$g_{P_o}(x) = -\ln(|1-2x|)\, I_{[0,1]}(x).$$

Thus (2.9.84) and (2.9.85) are valid when $C(x,y) = C_o(x,y) = xy$. Also, if $C(x,y) \geq xy$, then $P_o(C) \geq P_o(C_o)$ so that

$$P(P_o(C) \leq \delta(t)) \leq P(P_o(C_o) \leq \delta(t)) \quad (2.9.86)$$

Thus, using the exact calculations based on the *independence case*, it follows that

$$\forall\, C \geq xy,\ P(P_o(C) \leq \delta(t)) = 0(t),$$

We believe that when T and U are dependent, (2.9.85) is also true. The class of distributions for which the assumption (2.9.47) is not vacuous for any $d > 1$ needs to be examined.

After this proof of Theorem 2.9.2 was completed, we discussed the Poisson convergence problem with Persi Diaconis. He communicated the problem to Charles Stein. In his Neyman lecture at the IMS 1984 Annual meeting, Professor Stein outlined an alternative proof of the Poisson convergence using his methodology for approximation of probabilities (see Stein 1986). However, no rigorous version of Stein's proof has been published yet. The proof of Theorem 2.9.2, via the simple methodology used in proving Theorem 2.9.1, is still an open problem.

2.10 Matching variables with independent errors model.

Kadane (1978) proposes a normal model with measurement errors in the matching variables for situations in which two files of data pertain to same individuals and some matching variables are available, albeit with measurement error.

Suppose that originally, the triplets $(\mathbf{X}_i, \mathbf{Y}_i, \mathbf{Z}_i)$, $i = 1, 2, \ldots, n$ were a random sample from a multivariate normal population with mean vector $\underset{\sim}{\mu}$ and some covariance matrix Σ. It is assumed that when the data was broken into $(\mathbf{X}_i, \mathbf{Z}_i)$ and $(\mathbf{Y}_i, \mathbf{Z}_i)$ components, a measurement error $(\underset{\sim}{\varepsilon}_i, \underset{\sim}{\eta}_i)$ was made while recording variable $\mathbf{Z}_i$ in files 1 and 2. Therefore, the data in file 1 correspond to some unknown permutation of $(\mathbf{X}_i, \mathbf{Z}_i + \underset{\sim}{\varepsilon}_i)$ $i = 1, 2, \ldots, n$ and those in file 2 correspond to another unknown permutation of $(\mathbf{Y}_i, \mathbf{Z}_i + \underset{\sim}{\eta}_i)$, $i = 1, 2, \ldots, n$. It is also assumed by Kadane that $(\underset{\sim}{\varepsilon}_i, \underset{\sim}{\eta}_i)$, $i = 1, 2, \ldots, n$ are a random sample from a multivariate normal population with mean zero and covariance matrix Λ.

For this model, Kadane shows that finding the maximum likelihood matching strategy is again equivalent to solving a linear assignment problem. However, the method requires the knowledge of Σ_{XY}. Thus one either needs some prior information on Σ_{XY} or needs to make the assumption that $\mathbf{X}_i$ and $\mathbf{Y}_i$ are conditionally independent given $\mathbf{Z}_i + \underset{\sim}{\varepsilon}_i$ and $\mathbf{Z}_i + \underset{\sim}{\eta}_i$.

It is not clear whether the estimates of Σ_{XY} based on the merged file are any better than the ones based the prior information in the former case. In the latter case, since the model is completely specified all functionals of the parameters can be estimated from the two files separately. Thus the need for merging is not apparent.

Kadane (1978) provides some methods of assessing or imputing Σ_{XY} and other components of the covariance matrix.

Chapter 3. MERGING FILES OF DATA ON SIMILAR INDIVIDUALS

Problems of statistical matching in which the two micro-data files being matched consisted of the same individuals were discussed in Chapter 2. Furthermore, practical and legal reasons for the unavailability of micro data files on the same individuals were cited in Chapter 1. In such a situation, we may have two files of data that pertain to similar individuals. Allowing for some matching variables to be observed for each unit in the two files, we seek to merge the files so that inference problems relating to the variables not present in the same file can be addressed.

In this chapter, we first review the existing literature and then briefly discuss some alternatives to matching based on models in which the non-matching variables are conditionally independent given the values of the matching variables. Finally, we present the results of a Monte-Carlo study carried out to evaluate several statistical matching strategies for similar individuals.

3.1 Kadane's Matching Strategies for Multivariate Normal Models

Distance-based matching strategies were introduced in Section 1.5. One choice of distance measure in the matching methodology can be motivated by using a model in which the unobserved triplet $\mathbf{W} = (\mathbf{X}, \mathbf{Y}, \mathbf{Z})$ has a multivariate normal distribution. The set-up of the two files to be merged is as follows:

File 1 comprises a random sample of size n_1 on $(\mathbf{X}, \mathbf{Z})$, while File 2 consists of a random sample of size n_2 on $(\mathbf{Y}, \mathbf{Z})$. Furthermore, we expect very few or no records in the two files to correspond to the same individuals.

Statistically, this means that, for all practical purposes, the two random samples are themselves *independent*. We shall denote the sample data as follows:

$$\begin{aligned} &\text{File 1:} \qquad (\mathbf{X}_i, \mathbf{Z}_i),\ i = 1,2,\ldots,n_1 \text{ (Base)} \\ &\text{File 2:} \qquad (\mathbf{Y}_j, \mathbf{Z}_j),\ j = n_1+1,\ldots,n_1+n_2 \text{ (Supplementary).} \end{aligned} \tag{3.1.1}$$

Once completed, the matching process leads to more comprehensive synthetic files, namely

$$\begin{aligned} &\text{Synthetic File 1:} \qquad (\mathbf{X}_i, \mathbf{Y}_i^*, \mathbf{Z}_i), \qquad i = 1,2,\ldots,n_1 \\ &\text{Synthetic File 2:} \qquad (\mathbf{X}_j^*, \mathbf{Y}_j, \mathbf{Z}_j),\ j = n_1+1,\ldots,\ n_1+n_2. \end{aligned} \tag{3.1.2}$$

where, $\mathbf{Y}_i^*$ is an imputed value of $\mathbf{Y}$ that comes from the File 2 and $\mathbf{X}_j^*$ is an imputed value of $\mathbf{X}$ that is taken from the File 1, both by means of some matching strategy. We shall now review the matching methodology suggested by Kadane (1978).

Suppose that $\mathbf{W} = (\mathbf{X}, \mathbf{Y}, \mathbf{Z})$ has a multivariate normal distribution with mean vector $(\underset{\sim}{\mu}_x, \underset{\sim}{\mu}_y, \underset{\sim}{\mu}_z)$ and variance-covariance matrix

$$\Sigma = \begin{bmatrix} \Sigma_{xx} & \Sigma_{xy} & \Sigma_{xz} \\ \Sigma_{yx} & \Sigma_{yy} & \Sigma_{yz} \\ \Sigma_{zx} & \Sigma_{zy} & \Sigma_{zz} \end{bmatrix}.$$

The parameters $\Sigma_{xx}, \Sigma_{xz}, \Sigma_{yy}, \Sigma_{yz}, \Sigma_{zz}$ can all be estimated consistently using the marginal information on $(\mathbf{X}, \mathbf{Z})$ and $(\mathbf{Y}, \mathbf{Z})$ respectively in the two files. However, Σ_{xy} is an unidentified parameter, because the joint likelihood of the data on $(\mathbf{X}, \mathbf{Z})$ and $(\mathbf{Y}, \mathbf{Z})$ is free of the matrix Σ_{xy}. In fact, if Σ_{xy} is such that the matrix $\begin{bmatrix} \Sigma_{xx} & \Sigma_{xy} \\ \Sigma_{yx} & \Sigma_{yy} \end{bmatrix}$ is positive definite, nothing is learned from the data about Σ_{xy}, except in a Bayesian framework, where Σ_{xy}, Σ_{xz}, Σ_{yz} are, a priori, dependent. Even in this situation, the posterior distribution of Σ_{xy} is updated only through the updatings of the distributions of Σ_{xz} and Σ_{yz}. Kadane's approach to merging File 1 and File 2 consists of the following steps:

(i) Start with an imputed value of Σ_{xy} via some a priori distribution on the covariance matrix Σ, (ii) Complete Files 1 and 2 by predicting the missing data, **X** or **Y**, using the marginal information in the files, (iii) Match these "completed" files based on a distance measure between records of the two files, (iv) Estimate parameters such as

$$\gamma = \int g(\mathbf{w})\, dF(\mathbf{w}), \tag{3.1.3}$$

using the synthetic file resulting from Step (iii) and repeating the Steps (ii) through (iv) many times to find the sensitivity of the estimates to the imputed value of Σ_{xy}. Finally the results are weighted by using the apriori distribution on Σ. Further details of the steps outlined above are as follows:

Suppose that an imputed value of Σ_{xy} is available. Then we can assume that Σ_{xy} is known and complete the two files by means of conditional expectations. Let $\Sigma_{ab.c}$, for any letters a, b and c, be given by

$$\Sigma_{ab.c} = \Sigma_{ab} - \Sigma_{ac}\, \Sigma_{cc}^{-1}\, \Sigma_{cb} \tag{3.1.4}$$

Then the predicted value $\hat{\mathbf{Y}}$, of a missing **Y** in File 1 is given by

$$\hat{\mathbf{Y}} = E(\mathbf{Y}|\mathbf{X},\mathbf{Z})$$

$$= \underset{\sim}{\mu}_y + \Sigma_{yx.z}\, \Sigma_{xx.z}^{-1}\, (\underset{\sim}{x}-\underset{\sim}{\mu}_x) + \Sigma_{yz.x}\, \Sigma_{zz.x}^{-1}\, (\underset{\sim}{z}-\underset{\sim}{\mu}_z). \tag{3.1.5}$$

Similarly, the predicted value $\hat{\mathbf{X}}$, of a missing **X** in File 2 is given by

$$\hat{X} = E(X|Y,Z)$$

$$= \underset{\sim}{\mu}_x + \Sigma_{xy.z}\,\Sigma^{-1}_{yy.z}\,(Y-\underset{\sim}{\mu}_y) + \Sigma_{xy.y}\,\Sigma^{-1}_{zz.y}\,(z-\underset{\sim}{\mu}_z)\,. \qquad (3.1.6)$$

Using (3.1.3), (3.1.5) and (3.1.6), it is now easy to show that $\hat{T}_i = (X_i, \hat{Y}_i, Z_i)$, $i = 2,\ldots,n_1$ are independently and identically distributed as multivariate normal with mean vector $(\underset{\sim}{\mu}_x, \underset{\sim}{\mu}_y, \underset{\sim}{\mu}_z)$ and variance-covariance matrix

$$\Omega_1 = \begin{bmatrix} \Sigma_{xx} & \Lambda_1' & \Sigma_{xz} \\ \Lambda_1 & \Lambda_3 & \Lambda_2' \\ \Sigma_{zx} & \Lambda_2 & \Sigma_{zz} \end{bmatrix} \qquad (3.1.7)$$

where $\Lambda_1 = \Sigma_{yx.z}\,\Sigma^{-1}_{xx.z}\,\Sigma_{xx} + \Sigma_{yz.x}\,\Sigma^{-1}_{zz.x}\,\Sigma_{zx}$,

$$\Lambda_2 = \Sigma_{zx}\,\Sigma^{-1}_{xx.z}\,\Sigma_{xy.z} + \Sigma_{zz}\,\Sigma^{-1}_{zz.x}\,\Sigma_{zy.x},$$

and

$$\Lambda_3 = \Sigma_{yx.z}\,\Sigma^{-1}_{xx.z}\,\Sigma_{xx}\,\Sigma^{-1}_{xx.z}\,\Sigma_{xy.z} + \Sigma_{yz.x}\,\Sigma^{-1}_{zz.x}\,\Sigma_{zz}\,\Sigma^{-1}_{zz.x}\,\Sigma_{zy.x}$$

$$+ 2\Sigma_{yx.z}\,\Sigma^{-1}_{xx.z}\,\Sigma_{xz}\,\Sigma^{-1}_{zz.x}\,\Sigma_{zy.x}\,.$$

Also, the vectors $\hat{U}_j = (\hat{X}_j, Y_j, Z_j)$, $j = n_1+1,\ldots, n_1+n_2$, are independently distributed and have a common multivariate normal distribution with mean vector $(\underset{\sim}{\mu}_x, \underset{\sim}{\mu}_y, \underset{\sim}{\mu}_z)$ and variance-covariance matrix

$$\Omega_2 = \begin{bmatrix} \Lambda_4 & \Lambda_5' & \Lambda_6' \\ \Lambda_5 & \Sigma_{yy} & \Sigma_{yz} \\ \Lambda_6 & \Sigma_{zy} & \Sigma_{zz} \end{bmatrix}, \qquad (3.1.8)$$

where $\Lambda_4 = \Sigma_{xy.z}\,\Sigma^{-1}_{yy.z}\,\Sigma_{yy}\,\Sigma^{-1}_{yy.z}\,\Sigma_{yx.z}$

$$+ \Sigma_{xz.y} \Sigma^{-1}_{zz.y} \Sigma_{zz} \Sigma^{-1}_{zz.y} \Sigma_{zx.y} + 2\Sigma_{xy.z} \Sigma^{-1}_{yy.z} \Sigma_{yz} \Sigma^{-1}_{zz.y} \Sigma_{zx.y} ,$$

$$\Lambda_5 = \Sigma_{yy} \Sigma^{-1}_{yy.z} \Sigma_{yx.z} + \Sigma_{yz}\Sigma^{-1}_{zz.y} \Sigma_{zx.y} ,$$

and

$$\Lambda_6 = \Sigma_{zy} \Sigma^{-1}_{yy.z} \Sigma_{yx.z} + \Sigma_{zz}\Sigma^{-1}_{zz.y} \Sigma_{zx.y} .$$

Note that the distributions given by (3.1.7) and (3.1.8) are singular because the predicted values $\hat{Y}_i$ and $\hat{X}_{j+n_1}$ are linear functions of the other components of the random vectors $\hat{T}_i$ and $\hat{U}_j$ respectively. In order to describe Kadane's procedures to match the completed File 1, namely, $\hat{T}_1,...,\hat{T}_{n_1}$ with the completed File 2, namely, $\hat{U}_1,...,\hat{U}_{n_2}$, let us assume for simplicity that $n_1 = n_2 = n$. Starting with n records in each file, compute the differences $\{\hat{T}_i - \hat{U}_j\}$, for each $(i,j) \in \{1,...,n\}$in order to define a measure of dissimilarity between records from the two completed files. Now, $\hat{T}_i - \hat{U}_j$, $1 \leq i, j \leq n$ are identically distributed, multivariate normal random variables with mean **0** and covariance matrix $\Omega_1 + \Omega_2$, which can be shown to be *nonsingular*. For any positive semi-definite matrix A, a dissimilarity measure between the ith record of completed File 1 and jth record of completed File 2 can be defined by

$$d_{ij}(A) = (\hat{T}_i - \hat{U}_j)'A(\hat{T}_i - \hat{U}_j). \qquad (3.1.9)$$

Various choices of positive semi-definite matrix A in (3.1.9) provide different distance measures.

It may be recalled from Section 1.5 that a constrained matching of the two files is obtained by minimizing

$$C = \sum_{i=1}^{n} \sum_{j=1}^{n} d_{ij} a_{ij} \qquad (3.1.10)$$

subject to the conditions

$$\sum_{j=1}^{n} a_{ij} = 1, \quad \forall\, i = 1,2,\ldots,n \tag{3.1.11}$$

$$\sum_{i=1}^{n} a_{ij} = 1, \quad \forall\, j = 1,2,\ldots,n \tag{3.1.12}$$

and

$$a_{ij} = 0 \text{ or } 1, \forall \text{ i and j.} \tag{3.1.13}$$

If the d_{ij}'s are given by (3.1.9) for some choice of A, then distance-based optimal constrained match can be obtained by solving a linear assignment problem. Sometimes, an optimal matching may be obtained by minimizing (3.1.10) without requiring that the constraints (3.1.11) and (3.1.12) hold. However, as reported in Rodgers (1984), unconstrained optimal matches do not provide good estimates of the distribution $\mathbf{W} = (\mathbf{X}, \mathbf{Y}, \mathbf{Z})$.

It is important to note that the aforementioned optimization problem needs to be solved for each pair of base and supplementary files. If $\hat{\mathbf{T}}_i$ from File 1 and $\hat{\mathbf{U}}_j$ from File 2 have been matched in a given problem, then it might be natural to consider $(\mathbf{X}_i, \mathbf{Y}_j, \mathbf{Z}_i)$ and $(\mathbf{X}_i, \mathbf{Y}_j, \mathbf{Z}_j)$ as simulations from the underlying distribution. Now, the parameter γ in (3.1.3) can be estimated using one of the following synthetic samples:

$$\text{Synthetic File 1: } (\mathbf{X}_i, \mathbf{Y}_i^*, \mathbf{Z}_i,) \quad i = 1,2,\ldots,n, \tag{3.1.14}$$

$$\text{Synthetic File 2: } (\mathbf{X}_j^*, \mathbf{Y}_j, \mathbf{Z}_j), \quad j = n+1,\ldots,2n, \tag{3.1.15}$$

where $\mathbf{Y}_i^*$ and $\mathbf{X}_j^*$ are values given by the matching procedure.

Kadane suggests that matchings based on a fixed A in (3.1.9) and the consequent inferences based on synthetic files such as (3.1.14) or (3.1.15) must be repeated many times and the results must be averaged in some sensible way in order to explore the sensitivity of our findings to the initial imputed value of

Σ_{xy}. We have not pursued issues such as the actual choice of a prior on Σ and the sensitivity studies of inferences based on synthetic data. We shall now discuss Kadane's suggestions about the matrix A, which will be used in our Monte-Carlo study.

Kadane (1978) advocates two choices for the matrix A in the definition of distance measure d_{ij} in (3.1.9). The first one is

$$A_1 = (\Omega_1 + \Omega_2)^{-1}, \tag{3.1.16}$$

where Ω_1 and Ω_2 are the matrices in (3.1.7) and (3.1.8). This choice of A leads to the so-called *Mahalanobis distance* between the completed records in the two files. The second one

$$A_2 = \begin{bmatrix} 0 & 0 & 0 \\ 0 & 0 & 0 \\ 0 & 0 & \Sigma_{zz}^{-1} \end{bmatrix}, \tag{3.1.17}$$

provides the *bias-avoiding distance* function.

In general, the relative benefits of these two distance measures are not well understood, although the empirical studies by Barr et al (1982) and others reported in Rodgers (1984), indicate that the Mahalanobis distance strategy is worse than the one based on the bias-avoiding distance, in that the multivariate relationships among the variables **X**, **Y** and **Z** are less distorted when A_2 is used. Note that when A_2 is used, we need not complete the records in each file, since only the matching variables are used in the distance computation. A special case of (3.1.17) when **Z** has only one component will be discussed in the next subsection.

3.1.1 Isotonic Matching Strategy

In Section 3.3, we shall evaluate Kadane's matching strategies in the simple case when the triple **W** = (**X**, **Y**, **Z**) has a trivariate normal distribution. In order to facilitate such evaluations, we now show that, given only one matching variable, the matching strategy based on (3.1.17) can be implemented without using any algorithm to minimize distances.

For a scalar Z, using (3.1.17) in the distance function (3.1.9), the objective function (3.1.10) is equivalent to

$$C = \sum_{i=1}^{n} \sum_{j=1}^{n} (Z_{1i} - Z_{2j})^2 a_{ij}. \qquad (3.1.18)$$

In a constrained match, a_{ij}'s are subject to the constraints (3.1.11) to (3.1.13). Thus, (3.1.18) further simplifies to

$$\sum_{i=1}^{n} Z_{1i}^2 + \sum_{j=1}^{n} Z_{2j}^2 - 2 \sum_{i=1}^{n} \sum_{j=1}^{n} Z_{1i} Z_{2j} a_{ij} .$$

Hence, the minimization of distances reduces to maximizing

$$C^* = \sum_{i=1}^{n} \sum_{j=1}^{n} a_{ij} Z_{1i} Z_{2j} \qquad (3.1.19)$$

subject to the conditions (3.1.11) to (3.1.13) on the a_{ij}'s.

DeGroot and Goel (1976) show that, given the values z_{1i}'s and z_{2j}'s, the constrained maximization of C^* is equivalent to maximizing $\sum_{i=1}^{n} z_{1i} z_{2\varphi(i)}$ over all permutations φ of the integers 1,2,...,n. However, this latter extremal problem was encountered in Section 2.4, where the M.L.P. φ^* for certain bivariate matching problems was discussed. It follows that, with regard to the distance measure given by (3.1.17), where Z is scalar, the optimal matching strategy is to order the Z-values in the two files separately and then match the ith largest Z in File 1 with the ith largest Z in File 2. This explicit solution means that, if the matrix A_2 in equation (3.1.17) is used to minimize distances between records of the two files, then the synthetic file 1 is obtained by matching the **X**-concomitant of the ith order-statistic among Z's in File 1 with the **Y**-concomitant of the ith order statistic among Z's in File 2. We shall refer to this strategy as ***isotonic matching*** of the two files because the matching procedure is determined by the order-statistic of the Z's in File 1 and the order-statistics of the Z's in File 2.

3.1.2 Sims' Matching Strategy

In the preceding subsection, it was shown that one of the two matching strategies can be simplified to the point of not using any optimization algorithm in the matching procedure. Such simplification is clearly not possible when the triple (**X**, **Y**, **Z**) has a multidimensional **Z**. The whole idea of generating very large synthetic data sets by actually minimizing a sum of distances over all potential matches seems profligate from the computational view point. An alternative to distance-based strategies, suggested by Sims (1978), will now be outlined.

Sims has stressed the importance of exploiting the local sparseness or denseness of the sample data on the matching variables **Z**. A dense region of the **Z**-space is one within which we expect that the distributions of **X** and **Y** given **Z** change little. It is, at the same time, a region within which we have many observations. Sims has suggested that, within a dense region any arbitrary matching procedure will produce results that do not distort the joint distribution of **X**, **Y** and **Z**. Sparse **Z**-regions have few observations and statistical matching becomes difficult within a sparse region. Sims felt that in a sparse region, statistical matchings will almost certainly distort the joint distribution of **X**, **Y** and **Z**. He suggested that, in such a region, we should either not match at all or go beyond matching to more elaborate methods of generating synthetic data. However, Sims did not spell out any specific alternative to matching within sparse **Z**-regions.

In our Monte-Carlo Study for comparing Kadane's strategies with Sim's, which will be presented in Section 3.3, we created ten bins in the Z-space, namely $(-\infty, -1.00]$, $(-1.00, -0.75]$, $(-0.75, -0.50]$, $(-0.50, -0.25]$, $(-0.25, 0.00]$, $(0.00, 0.25]$, $(0.25, 0.50]$, $(0.50, 0.75]$, $(0.75, 1.00]$, $(1.0, +\infty)$. The conditional mean of X or Y, given Z did not change much inside the eight bins which were between -1.00 and 1.00. Hence, these latter bins were considered dense bins and the two bins in the left and right tail of the distribution of z were considered sparse bins. Within each dense bin, records of the two files were *randomly* matched, whereas the isotonic matching strategy of Subsection 3.1.1 was used in the sparse bins.

3.2 Alternatives to Statistical Matching Under Conditional Independence

In Sections 1.4 and 1.5, various matching techniques studied in the literature were discussed. One of them was based on matching records at the packet level. These strategies have been criticized in the file-merging literature on the grounds that such procedures implicitly make the assumption that the variables **X** and **Y** are conditionally independent, given the value of the overlapping variables **Z** [denoted by $\mathbf{X} \perp \mathbf{Y} \mid \mathbf{Z}$, following Dawid (1979)]. Such a critique was put forward by Sims (1972), and later by many others. It should be pointed out that this criticism is most appropriate when an observation $(\mathbf{X}, \mathbf{Z})$ is randomly matched with an observation $(\mathbf{Y}, \mathbf{Z})$ from among all equivalent $(\mathbf{Y}, \mathbf{Z})$'s, that form a packet, by achieving a minimum closeness score, as was done in Okner (1972).

Any statistical model for this type of matching uses an additional assumption that the data in File 1 is stochastically independent of the data in File 2. Clearly, such files of data cannot be used to test the validity of the first implicit assumption that $\mathbf{X} \perp \mathbf{Y} \mid \mathbf{Z}$. In addition, Sims (1978) has observed that matching itself for the purpose of estimating γ in (3.1.3) is unnecessary. He pointed out that, when $\mathbf{X} \perp \mathbf{Y} \mid \mathbf{Z}$ holds, one can write

$$dF(w) = dF^{XZ}(w)\, dF^{YZ}(w)/dF^{Z}(w), \qquad (3.2.1)$$

where $F^{XZ}(.)$ is the marginal (with regard to **W**) CDF of **X** and **Z** and the other terms on the right-hand side of (3.2.1) are analogously defined marginal distribution functions. Thus, the two separate samples in (3.1.1) are adequate to estimate all the terms on the right-hand side of (3.2.1) and alternatives to matching can be investigated. With emphasis on estimating the covariances or correlations between **X** and **Y**, we shall first review a histogram-type alternative, suggested by Sims (1978).

Suppose that we form a grid in the **W** space and estimate the joint density of **W** by first counting the number of sample points in each cell of the **z** grid. Let i index **X**-categories, j index **Y**-categories and k index **Z**-categories. Let n_{ijk} be the number of sample points in the $(i,j,k)^{th}$ cell. Using the dot notation to define counts of sample points with regard to marginal distributions, define $n^{(1)}_{i.k}$ = the number of sample points in File 1 with **X** in the ith category and **Z** in the kth category, $n^{(1)}_{.jk}$ = the number of sample points with **Y** in the jth category and **Z** in the kth category, and $n_{..k}$ = the number of sample points in Files 1 and 2 with **Z** in the kth category, i.e., $n_{..k} = \sum_i n^{(1)}_{i.k} + \sum_j n^{(1)}_{.jk}$.

The data in the two files given by (3.1.1) can be used to compute $n_{i.k}$, $n_{.jk}$ and $n_{..k}$ for all possible values of i, j and k. Thus, $n_{i.k}$ is obtained from File 1, $n_{.jk}$ from File 2 and $n_{..k}$ from the two files together. Finally, for a given function g, let $g(w_{ijk})$ denote the value of g computed at the center w_{ijk} of the $(i,j,k)^{th}$ cell in this grid. Sims has suggested that one could estimate γ in (3.1.3) by the statistic

$$\hat{\gamma} = \sum_{i,j,k} g(W_{ijk}) \frac{n^{(1)}_{i.k}\, n^{(2)}_{.jk}}{n_{..k}\, n_{...}} \tag{3.2.2}$$

With regard to $\hat{\gamma}$ in (3.2.2), theoretical properties such as the asymptotic distribution of $\hat{\gamma}$ (as the sample size tends to ∞) are unknown at the present time. Also, practical problems such as the choice of **W**-grid, and the cells thereof, which would keep the number of terms in the sum (3.2.2) computationally reasonable, have not yet been studied.

Sims (1978) stated that a procedure like the one leading to $\hat{\gamma}$ in (3.2.2), which takes into account the implicit assumption of conditional independence underlying various matching methodologies, has the following advantages over matching to create a synthetic file such as (3.1.14):

(a) the procedure lends itself to computation of standard errors indicating the reliability of estimators,

(b) the procedure can be connected to the large statistical literature on estimating density functions and multi-dimensional contingency tables, and

(c) it is likely to provide more accurate results than matching.

Given the lack of work on the statistical properties of the alternatives to matching, we seem to agree with the advantages (a) and (b), but regard (c) as an undemonstrated speculation. We shall not discuss $\hat{\gamma}$ in (3.2.2) any further. Nor shall we elaborate the merits and demerits of alternatives to matching and synthetic-data-based procedures. Nevertheless, in the next subsection, we shall derive the estimators of parameters for conditionally independent normal models without matching the files in (3.1.1).

3.2.1 ML Estimation in Multivariate Normal Models Using Two Files of Data

Consider the random vectors, **X**, **Y** and **Z**, with respective dimensions p, q and r. Suppose that **W** = (**X**, **Y**, **Z**) has a nonsingular multivariate normal distribution with unknown mean vector $(\underset{\sim}{\mu}_x, \underset{\sim}{\mu}_y, \underset{\sim}{\mu}_z)$ and unknown variance-covariance matrix Σ. Suppose that the sample data in (3.1.1) is available and that $n_1 \geq p + r$, $n_2 \geq q + r$.

In this section, we shall find the maximum likelihood estimators of the covariances among the variables in the vectors **X** and **Y** as well as other parameters, under the asumption that $\mathbf{X} \perp \mathbf{Y} \mid \mathbf{Z}$, without matching the files (3.1.1). The maximum likelihood estimation of parameters in multivariate normal models based on various patterns of missing data has been discussed in the literature. See, for example, Anderson (1984), Eaton and Kariya (1983), and

Srivastava and Khatri (1979). However, the pattern of data given by the set-up (3.1.1) does not seem to have been examined.

Note first that, under conditional independence, the density of **W** = (**X**, **Y**, **Z**) can be written as

$$f_W(\underset{\sim}{w};\theta) = f_1(\underset{\sim}{z};\theta)\, f_2(\underset{\sim}{x}|z, \theta)\, f_3(\underset{\sim}{y}|z, \theta) \tag{3.2.3}$$

where $\underset{\sim}{\theta} = (\underset{\sim}{\mu}_x, \underset{\sim}{\mu}_y, \underset{\sim}{\mu}_z, \Sigma_{xx}, \Sigma_{xy}, \Sigma_{xz}, \Sigma_{yy}, \Sigma_{yz}, \Sigma_{zz})$ and $f_1(.)$ is the marginal density function of **Z**, $f_2(.)$ and $f_3(.)$ are respectively the conditional densities of **X** and **Y**, given **Z** = **z**. It is well-known (Anderson, 1984, p. 33 and 37) that f_1, f_2 and f_3 also correspond to certain multivariate normal densities. Using the joint normality of **X**, **Y** and **Z**, it is also well known that (3.2.3) holds iff

$$\Sigma_{xy} = \Sigma_{xz}\, \Sigma_{zz}^{-1}\, \Sigma_{zy}\,. \tag{3.2.4}$$

It follows from (3.2.3) that the likelihood of $\underset{\sim}{\theta}$ given the observations in the two files in (3.1.1) is

$$L(\underset{\sim}{\theta}) = L_1(\underset{\sim}{\theta})\, L_2(\underset{\sim}{\theta})\, L_3(\underset{\sim}{\theta}), \tag{3.2.5}$$

where $$L_1(\underset{\sim}{\theta}) \propto \prod_{i=1}^{n_1+n_2} f_1(z_i, \underset{\sim}{\theta}), \tag{3.2.6}$$

$$L_2(\underset{\sim}{\theta}) \propto \prod_{i=1}^{n_1} f_2(x_i \mid z_i, \underset{\sim}{\theta}), \tag{3.2.7}$$

and $$L_3(\underset{\sim}{\theta}) \propto \prod_{i=n_1+1}^{n_1+n_2} f_3(y_i \mid z_i, \underset{\sim}{\theta}). \tag{3.2.8}$$

Taking natural logarithms of both sides of the equation (3.2.5), we obtain

$$\ell(\underset{\sim}{\theta}) = \sum_{k=1}^{3} \ell_k(\underset{\sim}{\theta}), \tag{3.2.9}$$

where $\quad \ell_k(\underset{\sim}{\theta}) = \ell n\,(L_k(\underset{\sim}{\theta})), \; k = 1, 2, 3.$

Let $\bar{\mathbf{Z}}$ and S_{zz} denote respectively the mean and the covariance matrix (with divisor $n_1 + n_2$) of the data $\mathbf{Z}_1,\ldots,\mathbf{Z}_{n_1+n_2}$, i.e.,

$$\bar{\mathbf{Z}} = \frac{1}{n_1 + n_2} \sum_{i=1}^{n_1+n_2} \mathbf{Z}_i \tag{3.2.10}$$

$$S_{zz} = \frac{1}{(n_1+n_2)} \sum_{i=1}^{n_1+n_2} (\mathbf{Z}_i - \bar{\mathbf{Z}})\;(\mathbf{Z}_i - \bar{\mathbf{Z}})'. \tag{3.2.11}$$

Similarly, let $\bar{\mathbf{Z}}_1$, $\bar{\mathbf{X}}$ and S_{xz}, S_{xx}, S_{zz1} denote the means and the covariance matrices (with divisor n_1) of (**X**, **Z**) data in File 1. Furthermore, let $\bar{\mathbf{Z}}_2$, $\bar{\mathbf{Y}}$, and S_{yz}, S_{yy}, S_{zz2} denote the means and covariance matrices (with divisor n_2) of (**Y**, **Z**) data in File 2.

The M.L.E. of $\underset{\sim}{\theta}$ to be obtained by maximizing $\ell(\underset{\sim}{\theta})$, in (3.2.9) over $\underset{\sim}{\theta}$, will be easier to derive if we reparametrize the distribution of **W**, and consequently the likelihood in (3.2.9) in terms of

$$\underset{\sim}{\eta} = (\underset{\sim}{\mu}_z, \underset{\sim}{\Sigma}_{zz}, \underset{\sim}{\nu}_{xz}\; \underset{\sim}{\nu}_{yz}, \Sigma_{xx.z}, \Sigma_{yy.z}, \mathbf{B}_{xz}, \mathbf{B}_{yz}), \tag{3.2.12}$$

where, $\Sigma_{ab.c}$ is defined in (3.1.4) and for arbitrary a and b

$$\mathbf{B}_{ab} = \Sigma_{ab}\,\Sigma_{bb}^{-1}, \quad \underset{\sim}{\nu}_{ab} = \underset{\sim}{\mu}_a - \mathbf{B}_{ab}\,\underset{\sim}{\mu}_b\,. \tag{3.2.13}$$

Since there is a one-to-one correspondence between $\underset{\sim}{\theta}$ and $\underset{\sim}{\eta}$, maximizing $L(\underset{\sim}{\theta})$ over $\underset{\sim}{\theta}$ is equivalent to maximizing $\ell_k(\underset{\sim}{\eta})$ over $\underset{\sim}{\eta}$, for each k = 1,2,3. The advantage of the transformation to the $\underset{\sim}{\eta}$-space is that $\ell_k(\underset{\sim}{\eta})$'s are functions of disjoint portions of $\underset{\sim}{\eta}$. In fact, $\ell_1(\underset{\sim}{\eta})$ is given by

$$\ell_1(\underset{\sim}{\eta}) = -\ln|\Sigma_{zz}| + \mathrm{tr}[\Sigma_{zz}^{-1}(s_{zz} + (\bar{\underset{\sim}{z}} - \mu_2)(\bar{\underset{\sim}{z}} - \mu_2)')] . \quad (3.2.14)$$

Furthermore, $\ell_2(\underset{\sim}{\eta})$ and $\ell_3(\underset{\sim}{\eta})$ are given by

$$-\ell_2(\underset{\sim}{\eta}) = \ln|\Sigma_{xx.z}| + \mathrm{tr}\left\{\Sigma_{xx.z}^{-1} M_{xz}\right\} , \quad (3.2.15)$$

$$-\ell_3(\underset{\sim}{\eta}) = \ln|\Sigma_{yy.z}| + \mathrm{tr}\left\{\underset{\sim}{\Sigma}_{yy.z}^{-1} M_{yz}\right\} , \quad (3.2.16)$$

where
$$M_{xz} = \left[\frac{1}{n_1}\sum_{i=1}^{n_1}(x_i - \underset{\sim}{\nu}_{xz} - B_{xz}z_i)(x_i - \underset{\sim}{\nu}_{xz} - B_{xz}z_i)'\right]$$

and
$$M_{yz} = \left[\frac{1}{n_2}\sum_{j=n_1+1}^{n_1+n_2}(y_j - \underset{\sim}{\nu}_{yz} - B_{yz}z_j)(y_j - \underset{\sim}{\nu}_{yz} - B_{yz}z_j)'\right] .$$

In view of Theorem 8.2.1 of Anderson (1984), and using (3.2.14) - (3.2.16), it can be easily shown, that M.L.E. of $\underset{\sim}{\eta}$ is given by

$$\hat{\underset{\sim}{\mu}}_z = \bar{Z}, \ \hat{\Sigma}_{zz} = S_{zz}, \quad (3.2.17)$$

$$\hat{B}_{xz} = S_{xz}S_{zz1}^{-1}, \ \hat{\underset{\sim}{\nu}}_{xz} = \bar{X} - \hat{B}_{xz}\bar{Z}_1 , \quad (3.2.18)$$

$$\hat{B}_{yz} = S_{yz}S_{zz2}^{-1}, \ \hat{\underset{\sim}{\nu}}_{yz} = \bar{Y} - \hat{B}_{yz}\bar{Z}_2 , \quad (3.2.19)$$

$$\hat{\Sigma}_{xx.z} = S_{xx} - S_{xz}S_{zz1}^{-1}S_{zx} , \quad (3.2.20)$$

and

$$\hat{\Sigma}_{yy.z} = S_{yy} - S_{yz}S_{zz2}^{-1}S_{zy} . \quad (3.2.21)$$

Using the estimators of $\underset{\sim}{\eta}$ in (3.2.17) - (3.2.21) and the relationships between $\underset{\sim}{\theta}$ and $\underset{\sim}{\eta}$, defined by (3.2.12) and (3.2.13), the M.L.E. of $\underset{\sim}{\theta}$ are given by the following equations:

$$\hat{\underset{\sim}{\mu}}_x = \bar{X} - \hat{B}_{xz} (\bar{Z}_1 - \bar{Z}), \; \hat{\underset{\sim}{\mu}}_y = \bar{Y} - \hat{B}_{yz} (\bar{Z}_2 - \bar{Z}), \tag{3.2.22}$$

$$\hat{\Sigma}_{xx} = S_{xx} - \frac{n_2}{n_1+n_2} \hat{B}_{xz} (S_{zz1} - S_{zz2}) \hat{B}'_{xz}, \; \hat{\Sigma}_{xz} = \hat{B}_{xz} S_{zz} \tag{3.2.23}$$

$$\hat{\Sigma}_{yy} = S_{yy} - \frac{n_1}{n_1+n_2} \hat{B}_{yz} (S_{zz2} - S_{zz1}) \hat{B}'_{yz}, \; \hat{\Sigma}_{yz} = \hat{B}_{yz} S_{zz}, \tag{3.2.24}$$

and finally,

$$\hat{\Sigma}_{xy} = S_{xz} S^{-1}_{zz1} S_{zz} S^{-1}_{zz2} S_{zy}, \tag{3.2.25}$$

where $\hat{B}_{xy}$ and $\hat{B}_{yz}$ are given by (3.2.18) and (3.2.19) respectively.

It follows from the above discussion that if the assumption that $\mathbf{X} \perp \mathbf{Y} \mid \mathbf{Z}$ is justifiable, then we can avoid matching the files in (3.1.1) and estimate Σ_{xy} as well as other parameters in the distribution of $\mathbf{W}$, by equations (3.2.17), (3.2.22) - (3.2.25). Unfortunately, the two data files contain no information regarding the appropriateness of this assumption, and prior information from other sources must be considered. The point here is that, if the matching strategy is based on assumptions like $\mathbf{X} \perp \mathbf{Y} \mid \mathbf{Z}$, then we must look for alternatives to matching whose statistical properties are known. Such alternatives may be useful especially because very little is known about the reliability of synthetic data-files.

It is important to note that if $\mathbf{X} \perp \mathbf{Y} \mid \mathbf{Z}$ holds and the appropriate moments of the distribution of $\mathbf{W}$ exist, then (3.2.4) holds even if $\mathbf{W}$ is not normally distributed. In any case, $\hat{\Sigma}_{xy}$ in (3.2.25) is a method of moments estimator and consequently it is strongly consistent for Σ_{xy}.

3.3 An Empirical Evaluation of Certain Matching Strategies

Several distance-based matching strategies for creating synthetic data have been discussed in Section 3.1. Specifically, two strategies due to Kadane (1978) and a strategy proposed by Sims (1978) were mentioned. In this section, we shall evaluate these three strategies via a Monte-Carlo study in the special case where $\mathbf{W} = (\mathbf{X}, \mathbf{Y}, \mathbf{Z})$, the unobservable vector, has a trivariate normal distribution. Before discussing the results of our study, we shall review earlier simulation studies of statistical matching procedures. A more comprehensive review of evaluations of statistical matching procedures can be found in Rodgers (1984).

Barr et al (1982) used, among others, a statistical model in which a vector $\mathbf{W} = (X, Y, Z_1, Z_2)$ has a four-dimensional normal distribution with zero means, unit variances and various levels of covariances among the four variables. Altogether, these investigators generated 100 pairs of independent files, namely File 1 comprising 200 observations on (X, Z_1, Z_2) and File 2 consisting of 200 observations on (Y, Z_1, Z_2) for each of 12 populations. The populations differed with respect to the covariances of the variables. Then, for each such pair of files, six statistical matches were performed, namely three constrained matches and three unconstrained matches. In each of these six matches, they used three distance functions. The first was a weighted sum of the absolute differences of the two Z variables between records of the two files and the other two were the Mahalanobis-distance and the bias-avoiding distance, which were discussed in Section 3.1. A summary of the findings of Barr et al (1982) is as follows.

All three distance measures provided accurate estimates of the variance of the Y variable when the constrained matching procedure was used. They also found that all three unconstrained matching procedures produced Y distributions that had means which were significantly different from the corresponding population values. The estimated covariances of Y with Z_1, Z_2, which were computed only for constrained matches, tended to be underestimates. With respect to the most important question in the context of merging files, namely the estimation of relationships between X and Y variables, it was reported that, if the conditional independence assumption was invalid, all statistical matching procedures provided extremely poor estimates of the covariance of X and Y. On the other hand, when the conditional independence assumption was valid, all six procedures provided generally quite accurate estimates of the X-Y covariance. Their simulations also indicated that the Mahalanobis distance measure produced less accurate matching than subjectively weighted distance measures.

As mentioned earlier, our own Monte-Carlo study is confined to a trivariate normal model with all parameters unknown. However, our findings are sufficiently interesting to justify their inclusion here. Assume that the following data are available for the purpose of estimating the unknown parameters of the normal distribution:

$$\text{File 1:} \qquad (X_i, Z_i), \; i = 1,2,\ldots,n \tag{3.3.1}$$

$$\text{File 2:} \qquad (Y_j, Z_j), \; j = n+1,\ldots,2n. \tag{3.3.2}$$

If the conditional independence assumption $X \perp Y \mid Z$, or equivalently,

$$\rho_{xy} = \rho_{xz}\,\rho_{yz} \tag{3.3.3}$$

holds, then all the parameters in the distribution of W can be estimated without merging the files in (3.3.1) and (3.3.2). In fact, it follows from (3.2.17) - (3.2.19) and (3.2.22) - (3.2.25) that the MLE of the parameters in the distribution of W, based on the data in Files 1 and 2 above, are given by

$$\hat{\mu} = \bar{Z},\; \hat{\mu}_x = \bar{X} - \tilde{\rho}_{xz}\frac{S_x}{S_{z1}}(\bar{Z}_1 - \bar{Z}_2),\;\; \hat{\mu}_y = \bar{Y} - \hat{\rho}_{yz}\frac{S_y}{S_{zz}}(\bar{Z}_2 - \bar{Z}_1), \tag{3.3.4}$$

$$\hat{\sigma}_Z^2 = S_Z^2,\;\; \hat{\sigma}_X^2 = S_X^2\, C_X^2,\;\; \hat{\sigma}_Y^2 = S_Y^2\, C_Y^2, \tag{3.3.5}$$

and

$$\hat{\rho}_{xz} = \tilde{\rho}_{xz} \left(\frac{S_z}{S_{z1}}\right) C_X^{-1}, \quad \hat{\rho}_{yz} = \tilde{\rho}_{yz} \left(\frac{S_z}{S_{z1}}\right) C_X^{-1}, \quad \hat{\rho}_{xy} = \hat{\rho}_{xz}\, \hat{\rho}_{yz}, \tag{3.3.6}$$

where

$$C_X^2 = \left[1 - \tilde{\rho}_{xz}^2 \left\{1 - \left(\frac{S_{z2}}{S_{z1}}\right)^2\right\}\right], \quad C_Y^2 = 1 - \tilde{\rho}_{yz}^2 \left\{1 - \left(\frac{S_{z1}}{S_{z2}}\right)^2\right\}, \tag{3.3.7}$$

and $\tilde{\rho}_{xz}$, $\tilde{\rho}_{yz}$ represent the Pearson-moment-correlation for (X, Z) data in File 1 and (Y, Z) in File 2 respectively. Note that if $\bar{Z}_1 \approx \bar{Z}_2 \approx \bar{Z}$, then $\hat{\mu}_x \approx \bar{X}$, and $\hat{\mu}_y \approx \bar{Y}$. Similarly, if $S_{z1} \approx S_{z2} \approx S_z$, then $C_x \approx 1$ and $C_y \approx 1$. In that case, $\hat{\sigma}_x^2 \approx S_x^2$, $\hat{\sigma}_y^2 = S_y^2$; $\hat{\rho}_{xz} \approx \tilde{\rho}_{xy}$, $\hat{\rho}_{yz} \approx \tilde{\rho}_{yz}$, and finally

$$\hat{\rho}_{xy} \approx \tilde{\rho}_{xz}\ \tilde{\rho}_{yz}. \tag{3.3.8}$$

For the sake of simplicity, we shall examine both the cases of conditional independence, with ρ_{xy} satisfying (3.3.3), and of conditional positive dependence, with ρ_{xy} satisfying

$$\rho_{xy} > \rho_{xz}\ \rho_{yz}. \tag{3.3.9}$$

Finally, we shall evaluate matching strategies only from the point of view of estimating ρ_{xy}, the correlation between variables which are *not* in the same file.

It is clear that, if the condition $X \perp Y \mid Z$ does not hold, then we should not estimate ρ_{xy} by means of (3.3.6). In such a case, matching the files (3.3.1) and (3.3.2) for estimation purposes is an alternative worth studying. Thus, if after merging, File 1 becomes the synthetic File 1 namely

$$(X_i, Y_i^*, Z_i), \quad i = 1,2,\ldots,n \tag{3.3.10}$$

where Y_i^* is the value of Y assigned to the ith record in the process of merging, then we shall use the synthetic data (X_i, Y_i^*), $i = 1,2,\ldots,n$ to estimate ρ_{xy}.

Our program for an empirical evaluation of matching strategies is as follows:

(i) Generate data from normal population of W, with means 0, Variances 1 and correlations ρ_{xy}, ρ_{xz}, ρ_{yz}, to create independent files (3.3.1) and (3.3.2). Note that data on (X, Y), which is typically missing in actual matching situations, is available in simulation studies.

(ii) Using any given matching strategy, merge the two files created in Step (i) and compute the "synthetic correlation", denoted by $\hat{\rho}_s$, which is defined to be the sample correlation coefficient based on the (X, Y*) data given by the synthetic file (3.3.10)

(iii) Compare $\hat{\rho}_s$ of Step (ii) with the following correlation estimators.

(a) $\hat{\rho}_{m\ell 1}$: the sample correlation coefficient based on the unbroken data (X_i, Y_i), i = 1,2,...,n which was generated in Step (i). Observe that, if there is no apriori restriction on the parameters, then $\hat{\rho}_{m\ell 1}$ is the maximum likelihood estimator of ρ_{xy},based on W data in step (i).

(b) $\hat{\rho}_{m\ell 2}$: the estimator of ρ_{xy} given by (3.3.8), which is also the approximate maximum likelihood estimator of ρ_{xy} when conditional independence holds.

Since $\hat{\rho}_{m\ell 1}$ and $\hat{\rho}_{m\ell 2}$, respectively, are based on one file containing (X, Y) and two files containing (X, Z) and (Y, Z), we shall also refer to these as one-sample and two-sample estimates of ρ_{xy}.

Using the above program, we shall evaluate Kadane's distance-based matching strategies discussed in Section 3.1, namely the isotonic matching (or equivalently, bias-avoiding distance matching) strategy and the procedure induced by the Mahalanobis distance, as well as the method of matching in bins, which is an adaptation of a strategy due to Sims (1978), as explained in Subsection 3.1.2. The synthetic correlations resulting from the use of these three strategies will be denoted by $\hat{\rho}_{s1}$ (Rhohat Isotonic), $\hat{\rho}_{s2}$ (Rhohat M/Nobis), and $\hat{\rho}_{s3}$ (Rhohat Bin Match) respectively.

Our study has been conducted for three values of n, namely 10, 25 and 50. The values of the population correlation ρ_{xy} which are used, among others, to generate random deviates from the normal population of **W** = (X, Y, Z), were chosen from the following categories;

$$\begin{aligned} &\text{Low } \rho_{xy}: && 0.00,\ 0.25 \\ &\text{Medium } \rho_{xy}: && 0.50,\ 0.60,\ 0.65,\ 0.70 \\ &\text{High } \rho_{xy}: && 0.75\ (0.05)\ 0.95,\ 0.99 \end{aligned} \tag{3.3.11}$$

Combined with low as well as high values of ρ_{xz} and ρ_{yz}, there were 15 choices of ρ_{xy} from (3.3.11) such that the conditional independence restriction (3.3.3) was satisfied.

As remarked earlier, these correlations where chosen mainly to provide a basis such that the estimates of ρ_{xy} resulting from the case of conditional positive dependence can be compared with those resulting from conditional independence. The fifteen values of ρ_{xy} in the conditional independence cases were increased in such a way that the positive dependence was achieved. Altogether, nineteen such sets of $(\rho_{xy}, \rho_{xz}, \rho_{yz})$ were selected in our study.

For n = 10, **W** was generated 1000 times by using the IMSL subroutines. The calculation of $\hat{\rho}_{S1}$ was based on matching by sorting Z's in the two files, as discussed in Section 3.1.1. Furthermore, $\hat{\rho}_{S2}$ was computed for each realization by matching based on solving a linear assignment problem, while using the actual value of ρ_{xy} as the imputed value in completing the files. The Ford-Fulkerson algorithm (Zionts, 1974) was used for this purpose. The computing cost for solving assignment problems grows quite rapidly with n. Therefore, only 700 independent samples of size n = 25 were generated. A comprehensive examination of the results for n = 10, 25, revealed that $\hat{\rho}_{s1}$ and $\hat{\rho}_{s2}$, the correlations corresponding to two distance measures, were, for all practical purposes, identical (see Figures for n = 10 and 25 among Figures 3.1-3.18 and Figure 3.40 given in Appendix B), when conditional independence was valid. In case of conditional positive dependence, the performance of these two strategies was identical for n = 10. For n = 25, the isotonic strategy provided somewhat better estimates (see Figure 3.41 and n = 10, 25 figures among Figures 3.19-

3.36) than the Mahalanobis strategy in some cases and vice-versa in others. In view of this and the high computational costs for M/Nobis strategy, we compared only two matching strategies, the isotonic and the method of matching in bins for n = 50 (2500 independent samples).

The empirical distribution functions (e.d.f.'s) were obtained for the simulated data on $\hat{\rho}_{m\ell 1}$, $\hat{\rho}_{m\ell 2}$, $\hat{\rho}_{s1}$, $\hat{\rho}_{s2}$, $\hat{\rho}_{s3}$, which were calculated for the 34 sets of correlations selected for the study. However, we provide these plots only for a representative collection of 13 of these sets in Figures 3.1-3.39. We have also included the scatter plots for $\hat{\rho}_{s2}$, $\hat{\rho}_{s3}$, $\hat{\rho}_{m\ell 1}$, $\hat{\rho}_{m\ell 2}$ versus $\hat{\rho}_{s1}$ in Figures 3.40-3.45 to provide an illustration of the relationships between these estimators.

3.3.1 Conclusion from the Monte-Carlo Study

We shall first discuss our results in the case of conditional independence. The figures for n = 10 and 25, among Figures 3.1-3.18, clearly show that the estimates $\hat{\rho}_{s1}$ and $\hat{\rho}_{s2}$ based on the isotonic matching strategy and Mahalanobis distance based strategy, respectively, have nearly identical e.d.f. Figure 3.40 provides the evidence of the fact that the two estimators are nearly identical. In fact, an examination of all the results showed that, for all values of n and Σ in our study, the estimates $\hat{\rho}_{s1}$ and $\hat{\rho}_{s2}$ were almost identical.

As noted in Section 3.2, $\hat{\rho}_{m\ell 2}$, the approximate maximum likelihood estimator of ρ_{xy} under this model, and $\hat{\rho}_{m\ell 1}$, the method of moments estimator based on paired-data, are computed for comparison purposes. As expected, $\hat{\rho}_{m\ell 1}$ and $\hat{\rho}_{m\ell 2}$ behave equally well on the average even though the variability of $\hat{\rho}_{m\ell 1}$, as measured by the interquartile range, is consistently higher than that of $\hat{\rho}_{m\ell 2}$. Furthermore the ranges of $\hat{\rho}_{m\ell 1}$ are consistently bigger than those of $\hat{\rho}_{m\ell 2}$ (see Figures 3.1-3.18).

For low ρ_{xy} and each n, $\hat{\rho}_{s1}$, $\hat{\rho}_{s2}$, $\hat{\rho}_{s3}$ and $\hat{\rho}_{m\ell 1}$ have almost identical e.d.f. The median of these estimates are the same as that of $\hat{\rho}_{m\ell 2}$. However, these estimators have much larger variability than $\hat{\rho}_{m\ell 2}$, as shown in Figure 01-03. Furthermore, there is no meaningful relationship between the synthetic data estimators and $\hat{\rho}_{m\ell 1}$, $\hat{\rho}_{m\ell 2}$ as shown in Figures 3.42-3.43. For medium and high values of ρ_{xy}, all three synthetic estimators exhibit negative bias, whereas both $\hat{\rho}_{m\ell 1}$ and $\hat{\rho}_{m\ell 2}$ seem to be unbiased. $\hat{\rho}_{s3}$, the estimator given by the method of matching in bins, is more negatively biased than $\hat{\rho}_{s1}$ and $\hat{\rho}_{s2}$. Figures 3.4-3.18 illustrate these points and the fact that $\hat{\rho}_{s3}$ is worse than $\hat{\rho}_{s1}$ and $\hat{\rho}_{s2}$. These patterns among the five estimates exist for each sample size even though the difference between synthetic data estimators and $\hat{\rho}_{m\ell 2}$ tends to decrease as n increases. Finally as expected, $\hat{\rho}_{m\ell 1}$ has a higher variability as compared to $\hat{\rho}_{m\ell 2}$.

Turning to the conditional positive dependence case, we first note that $\hat{\rho}_{m\ell 1}$ is a reasonable estimator of ρ_{xy}, even though it would not be available to the practitioner. Furthermore, for $n = 10$, two distance based strategies are similar, whereas for $n = 25$, each one is somewhat better than the other depending on the various Σ values (see Figure 3.41 and figures for $n = 10$ and 25 among Figures 3.19-3.36). On comparing $\hat{\rho}_{m\ell 1}$ with $\hat{\rho}_{m\ell 2}$ and the synthetic data estimators $\hat{\rho}_{s1}$, $\hat{\rho}_{s2}$, and $\hat{\rho}_{s3}$, we find that these estimators perform very badly, in that all of them are consistently underestimates and therefore heavily negatively biased (see Figures 3.19-3.36).

In general, the synthetic data estimators are either comparable to $\hat{\rho}_{m\ell 2}$, or have a definite negative bias compared with $\hat{\rho}_{m\ell 2}$. Thus they are worse than $\hat{\rho}_{m\ell 2}$. Figures 3.19-3.36 support this conclusion. Furthermore, we observe that $\hat{\rho}_{s3}$ based on binning, never performs better than $\hat{\rho}_{s1}$ or $\hat{\rho}_{s2}$ as illustrated by Figure 3.45 and Figures 3.19-3.36. However, the distances between e.d.f. of $\hat{\rho}_{m\ell 2}$ and $\hat{\rho}_{si}$, $i = 1,2,3$ tend to decrease as n increases.

Finally it must be pointed out that as the positive dependence increases; i.e., $\rho_{xy} - \rho_{xz}\rho_{yz}$ increases, the distances between the e.d.f. of $\hat{\rho}_{m\ell 2}$ and each of the three synthetic data estimators and $\hat{\rho}_{m\ell 1}$ increases. Figures 3.20, and 3.37-3.39 illustrate this assertion. In particular, the conclusion seems to be that $\hat{\rho}_{m\ell 1}$ has the best performance, $\hat{\rho}_{m\ell 2}$ is the second best, $\hat{\rho}_{s1}$ and $\hat{\rho}_{s2}$ are the next and $\hat{\rho}_{s3}$ is the worst.

Based on these observations, we must conclude that when conditional independence model holds, the synthetic data estimators do not provide any advantage over $\hat{\rho}_{m\ell 2}$, the no-matching estimator. In fact, they are slightly worse than the $\hat{\rho}_{m\ell 2}$. On the other hand, in the case of conditional positive dependence, $\hat{\rho}_{m\ell 2}$ and all the synthetic data estimators perform badly. The performance of synthetic data estimators is never better than that of $\hat{\rho}_{m\ell 2}$. Thus estimators based on matching strategies do not seem to provide any advantage over the estimators based on no matching, but under the assumption of conditional independence. Thus for estimating ρ_{xy} from files containing data on similar individuals, the extra work involved in matching data files is almost worthless unless one uses a strong prior information of conditional positive dependence and thus creates better synthetic files.

Further studies are in order for much larger sample sizes to examine if this picture changes at all. We should point out that it is possible that matching may be useful for extracting some other features of the joint distribution instead of estimating the correlation of non-overlapping variables. Additional large scale Monte Carlo studies are warranted to explore this.

APPENDIX A
Tables 2.1 - 2.8

Table 2.1 Expected Average Number of ε -correct Matchings (ε =.01)
[Yahav(1982)]

ρ	$\mu_{10}(\varepsilon)$	$\mu_{20}(\varepsilon)$	$\mu_{50}(\varepsilon)$	$\mu(\varepsilon)$
.001	.5864	.5326	.5275	.5227
.01	.1984	.1648	.1271	.1152
.10	.1512	.1058	.0760	.0591
.30	.1084	.0686	.0389	.0214
.50	.1020	.0582	.0272	.0138
.70	.0960	.0614	.0262	.0105
.90	.0972	.0540	.0206	.0086
.95	.0976	.0496	.0214	.0083
.99	.0960	.0484	.0213	.0080

Table 2.2 Expected Average number of ε-correct Matchings (ε = 0.01)

ρ	$\mu_{10}(\varepsilon)$	$\mu_{20}(\varepsilon)$	$\mu_{50}(\varepsilon)$	$\mu_{100}(\varepsilon)$	$\mu(\varepsilon)$
0.00	0.106	0.054	0.025	0.015	0.008
0.10	0.113	0.059	0.028	0.017	0.008
0.20	0.127	0.068	0.031	0.018	0.008
0.30	0.138	0.075	0.034	0.020	0.008
0.40	0.155	0.083	0.038	0.023	0.008
0.50	0.174	0.095	0.044	0.026	0.008
0.60	0.199	0.109	0.061	0.036	0.008
0.70	0.231	0.129	0.061	0.036	0.008
0.80	0.279	0.162	0.077	0.046	0.016
0.90	0.374	0.222	0.109	0.067	0.016
0.95	0.476	0.296	0.151	0.094	0.024
0.99	0.700	0.521	0.299	0.191	0.056

Table 2.3 Expected Average number of ε-correct Matchings ε =(0.05)

ρ	$\mu_{10}(\varepsilon)$	$\mu_{20}(\varepsilon)$	$\mu_{50}(\varepsilon)$	$\mu_{100}(\varepsilon)$	$\mu(\varepsilon)$
0.00	0.127	0.076	0.047	0.037	0.032
0.10	0.134	0.082	0.051	0.040	0.032
0.20	0.149	0.093	0.056	0.043	0.032
0.30	0.161	0.099	0.061	0.047	0.032
0.40	0.180	0.109	0.066	0.052	0.040
0.50	0.201	0.124	0.074	0.057	0.040
0.60	0.228	0.141	0.085	0.065	0.048
0.70	0.262	0.166	0.101	0.076	0.048
0.80	0.317	0.205	0.124	0.094	0.064
0.90	0.420	0.280	0.174	0.135	0.088
0.95	0.529	0.368	0.237	0.186	0.127
0.99	0.769	0.631	0.459	0.377	0.274

Table 2.4 Expected Average number of ε -correct Matchings (ε = 0.1)

ρ	$\mu_{10}(\varepsilon)$	$\mu_{20}(\varepsilon)$	$\mu_{50}(\varepsilon)$	$\mu_{100}(\varepsilon)$	$\mu(\varepsilon)$
0.00	0.154	0.102	0.075	0.065	0.056
0.10	0.160	0.110	0.080	0.069	0.056
0.20	0.177	0.121	0.087	0.074	0.064
0.30	0.189	0.130	0.093	0.080	0.064
0.40	0.210	0.143	0.101	0.088	0.072
0.50	0.234	0.161	0.112	0.096	0.080
0.60	0.264	0.181	0.127	0.108	0.088
0.70	0.302	0.210	0.149	0.126	0.103
0.80	0.363	0.258	0.182	0.154	0.127
0.90	0.477	0.347	0.254	0.218	0.174
0.95	0.594	0.452	0.342	0.299	0.251
0.99	0.839	0.744	0.630	0.580	0.522

Table 2.5 Expected Average number of ε -correct Matchings (ε = 0.3)

ρ	$\mu_{10}(\varepsilon)$	$\mu_{20}(\varepsilon)$	$\mu_{50}(\varepsilon)$	$\mu_{100}(\varepsilon)$	$\mu(\varepsilon)$
0.00	0.255	0.208	0.184	0.175	0.166
0.10	0.265	0.223	0.195	0.186	0.174
0.20	0.284	0.237	0.207	0.197	0.190
0.30	0.305	0.253	0.221	0.211	0.197
0.40	0.334	0.275	0.240	0.229	0.213
0.50	0.363	0.304	0.263	0.250	0.236
0.60	0.401	0.336	0.293	0.278	0.266
0.70	0.455	0.382	0.337	0.320	0.303
0.80	0.532	0.457	0.403	0.386	0.362
0.90	0.670	0.593	0.540	0.519	0.497
0.95	0.802	0.733	0.689	0.674	0.658
0.99	0.978	0.968	0.961	0.961	0.966

Table 2.6 Expected Average number of ε -correct Matchings (ε = 0.5)

ρ	$\mu_{10}(\varepsilon)$	$\mu_{20}(\varepsilon)$	$\mu_{50}(\varepsilon)$	$\mu_{100}(\varepsilon)$	$\mu(\varepsilon)$
0.00	0.353	0.311	0.290	0.281	0.274
0.10	0.363	0.330	0.306	0.298	0.289
0.20	0.390	0.348	0.325	0.315	0.311
0.30	0.417	0.371	0.344	0.336	0.326
0.40	0.452	0.400	0.373	0.362	0.354
0.50	0.485	0.437	0.404	0.393	0.383
0.60	0.528	0.478	0.446	0.435	0.425
0.70	0.591	0.536	0.506	0.495	0.484
0.80	0.675	0.628	0.594	0.584	0.570
0.90	0.811	0.773	0.752	0.744	0.737
0.95	0.917	0.896	0.888	0.885	0.886
0.99	0.998	0.999	0.999	0.999	1.000

Table 2.7 Expected Average number of ε-correct Matchings (ε=0.75)

ρ	$\mu_{10}(\varepsilon)$	$\mu_{20}(\varepsilon)$	$\mu_{50}(\varepsilon)$	$\mu_{100}(\varepsilon)$	$\mu(\varepsilon)$
0.00	0.468	0.433	0.416	0.409	0.404
0.10	0.488	0.454	0.437	0.429	0.425
0.20	0.514	0.477	0.461	0.453	0.445
0.30	0.539	0.505	0.487	0.480	0.471
0.40	0.582	0.542	0.522	0.514	0.503
0.50	0.621	0.586	0.560	0.555	0.547
0.60	0.662	0.633	0.613	0.606	0.59
0.70	0.727	0.694	0.679	0.673	0.668
0.80	0.810	0.786	0.772	0.768	0.766
0.90	0.919	0.908	0.906	0.904	0.907
0.95	0.979	0.976	0.978	0.979	0.982
0.99	1.000	1.000	1.000	1.000	1.000

Table 2.8 Expected Average number of ε -correct Matchings (ε = 1.0)

ρ	$\mu_{10}(\varepsilon)$	$\mu_{20}(\varepsilon)$	$\mu_{50}(\varepsilon)$	$\mu_{100}(\varepsilon)$	$\mu(\varepsilon)$
0.00	0.570	0.545	0.531	0.524	0.522
0.10	0.593	0.566	0.555	0.549	0.547
0.20	0.621	0.595	0.581	0.576	0.570
0.30	0.646	0.622	0.611	0.605	0.605
0.40	0.690	0.664	0.650	0.644	0.627
0.50	0.729	0.707	0.691	0.688	0.683
0.60	0.772	0.753	0.744	0.741	0.737
0.70	0.830	0.812	0.807	0.805	0.803
0.80	0.898	0.889	0.887	0.885	0.886
0.90	0.970	0.970	0.972	0.972	0.975
0.95	0.996	0.996	0.997	0.997	0.998
0.99	1.000	1.000	1.000	1.000	1.000

APPENDIX B

Figures 3.1 - 3.45

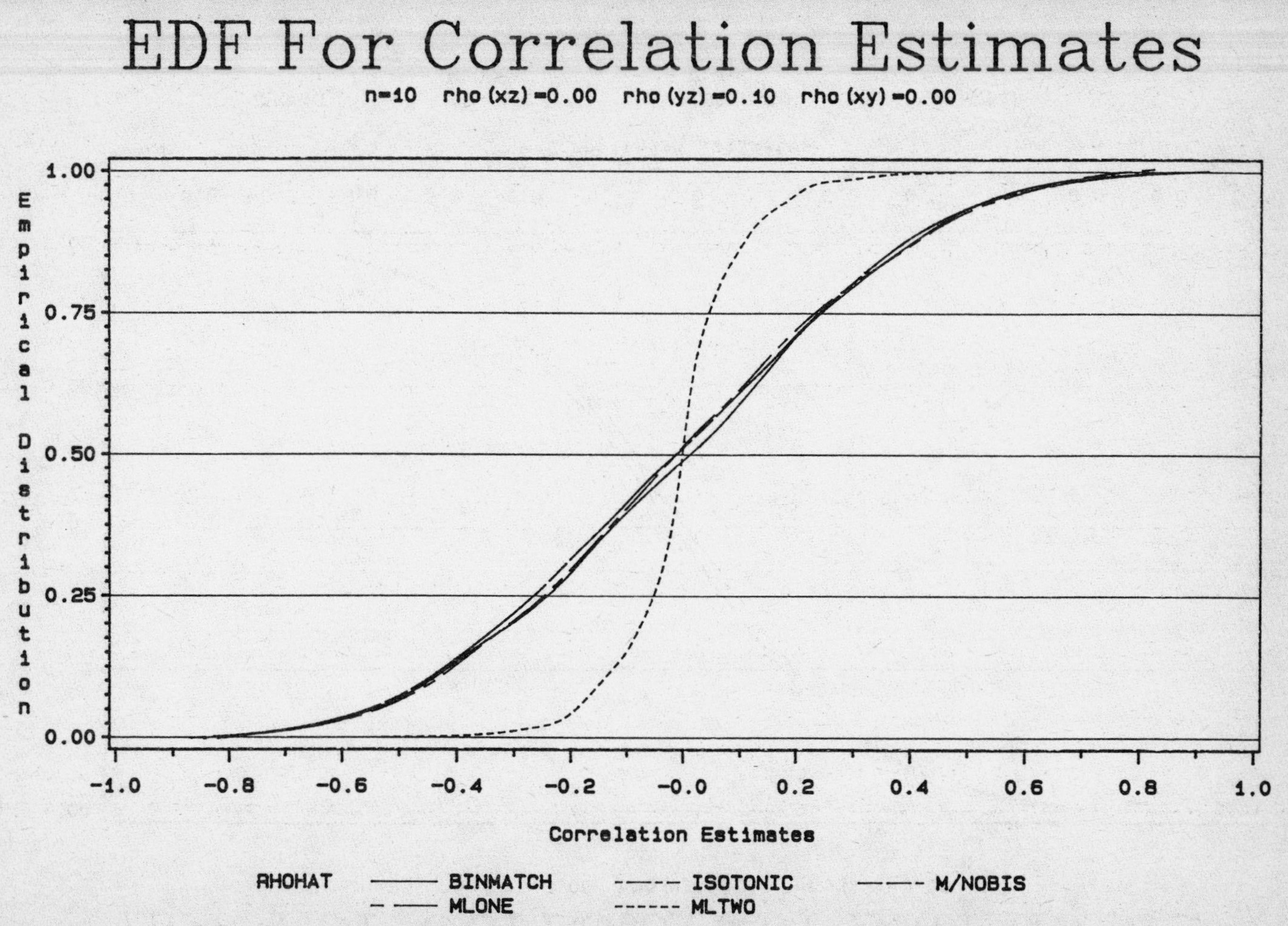
EDF For Correlation Estimates
n=10 rho (xz) =0.00 rho (yz) =0.10 rho (xy) =0.00
Empirical Distribution
1.00
0.75
0.50
0.25
0.00
-1.0
-0.8
-0.6
-0.4
-0.2
-0.0
0.2
0.4
0.6
0.8
1.0
Correlation Estimates
RHOHAT
BINMATCH
ISOTONIC
M/NOBIS
MLONE
MLTWO

Figure 3.1

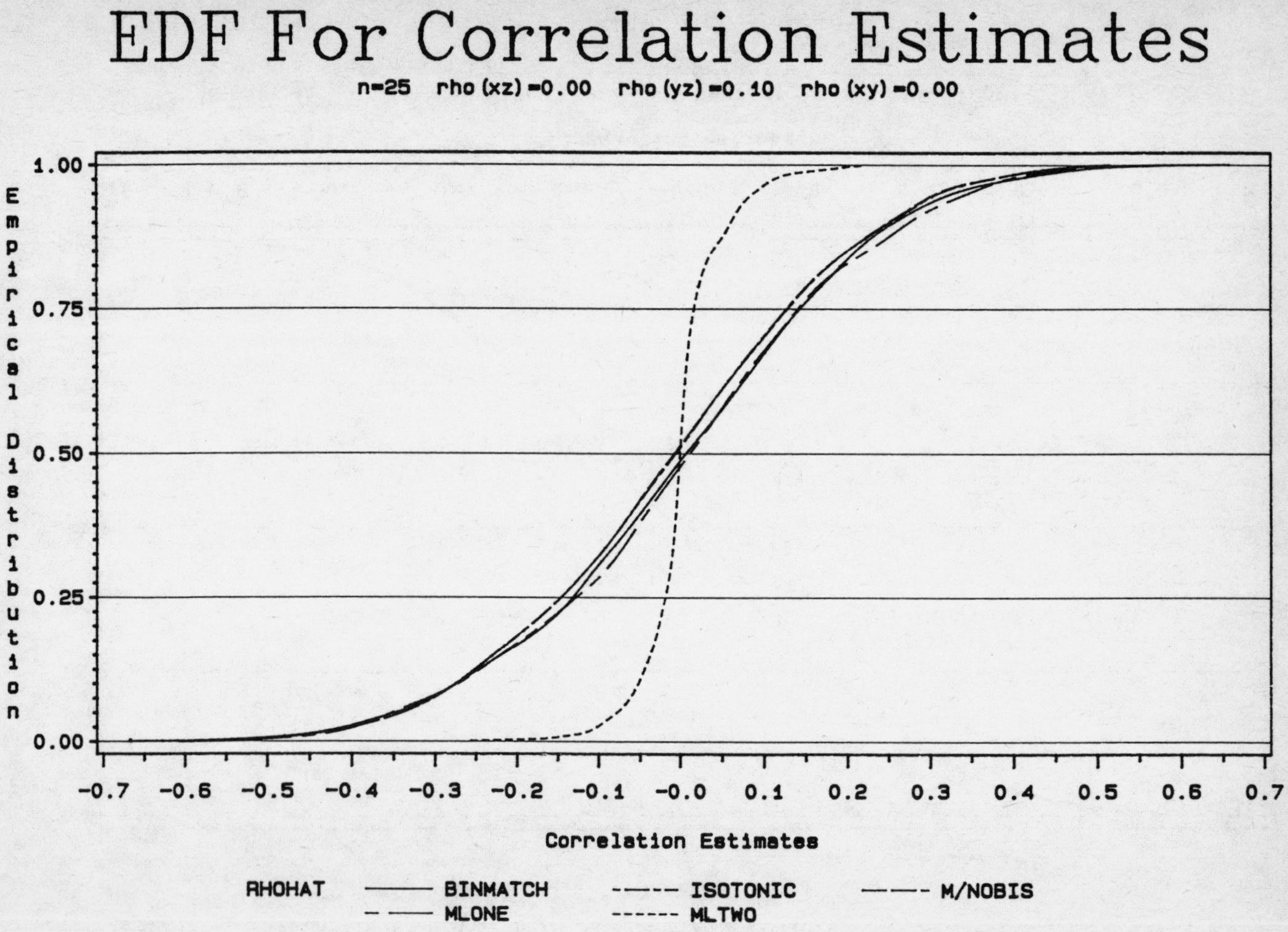
EDF For Correlation Estimates
n=25 rho (xz) =0.00 rho (yz) =0.10 rho (xy) =0.00
Empirical Distribution
1.00
0.75
0.50
0.25
0.00
-0.7 -0.6 -0.5 -0.4 -0.3 -0.2 -0.1 -0.0 0.1 0.2 0.3 0.4 0.5 0.6 0.7
Correlation Estimates
RHOHAT
BINMATCH
ISOTONIC
M/NOBIS
MLONE
MLTWO

Figure 3.2

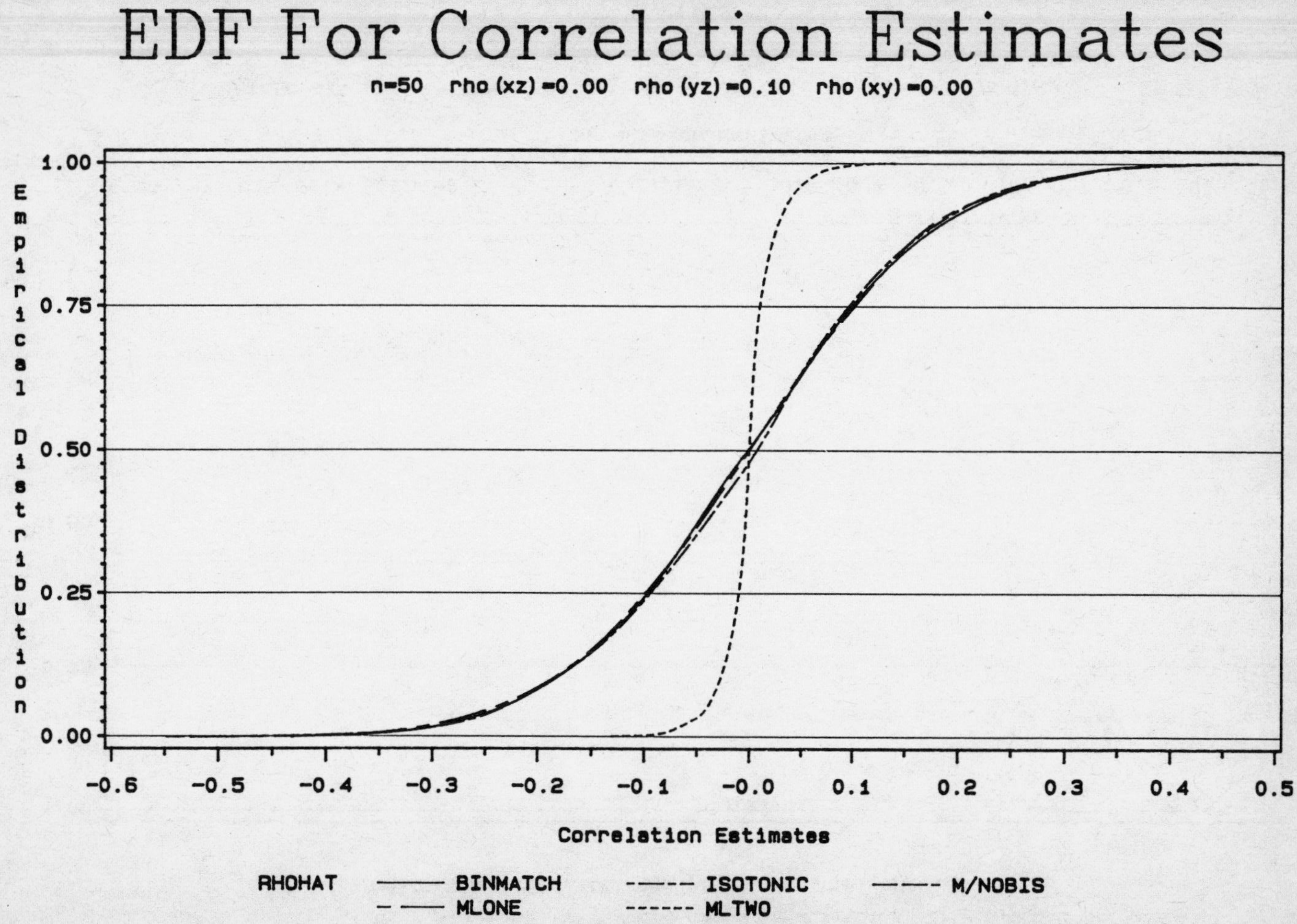
EDF For Correlation Estimates
n=50 rho (xz) =0.00 rho (yz) =0.10 rho (xy) =0.00
Empirical Distribution
1.00
0.75
0.50
0.25
0.00
-0.6
-0.5
-0.4
-0.3
-0.2
-0.1
-0.0
0.1
0.2
0.3
0.4
0.5
Correlation Estimates
RHOHAT
BINMATCH
ISOTONIC
M/NOBIS
MLONE
MLTWO

Figure 3.3

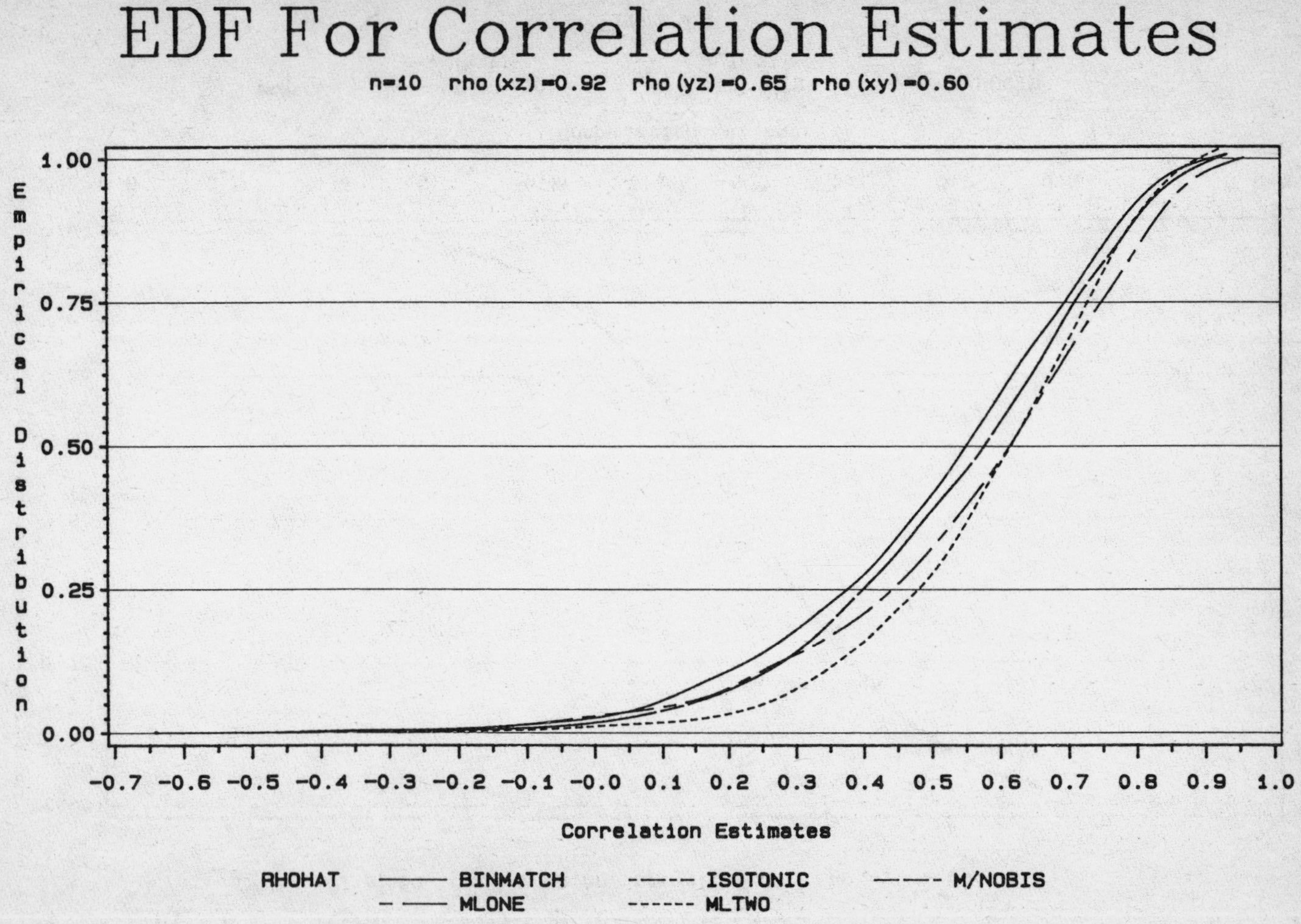
EDF For Correlation Estimates
n=10 rho (xz) =0.92 rho (yz) =0.65 rho (xy) =0.60
Empirical Distribution
1.00
0.75
0.50
0.25
0.00
-0.7 -0.6 -0.5 -0.4 -0.3 -0.2 -0.1 -0.0 0.1 0.2 0.3 0.4 0.5 0.6 0.7 0.8 0.9 1.0
Correlation Estimates
RHOHAT
BINMATCH
ISOTONIC
M/NOBIS
MLONE
MLTWO

Figure 3.4

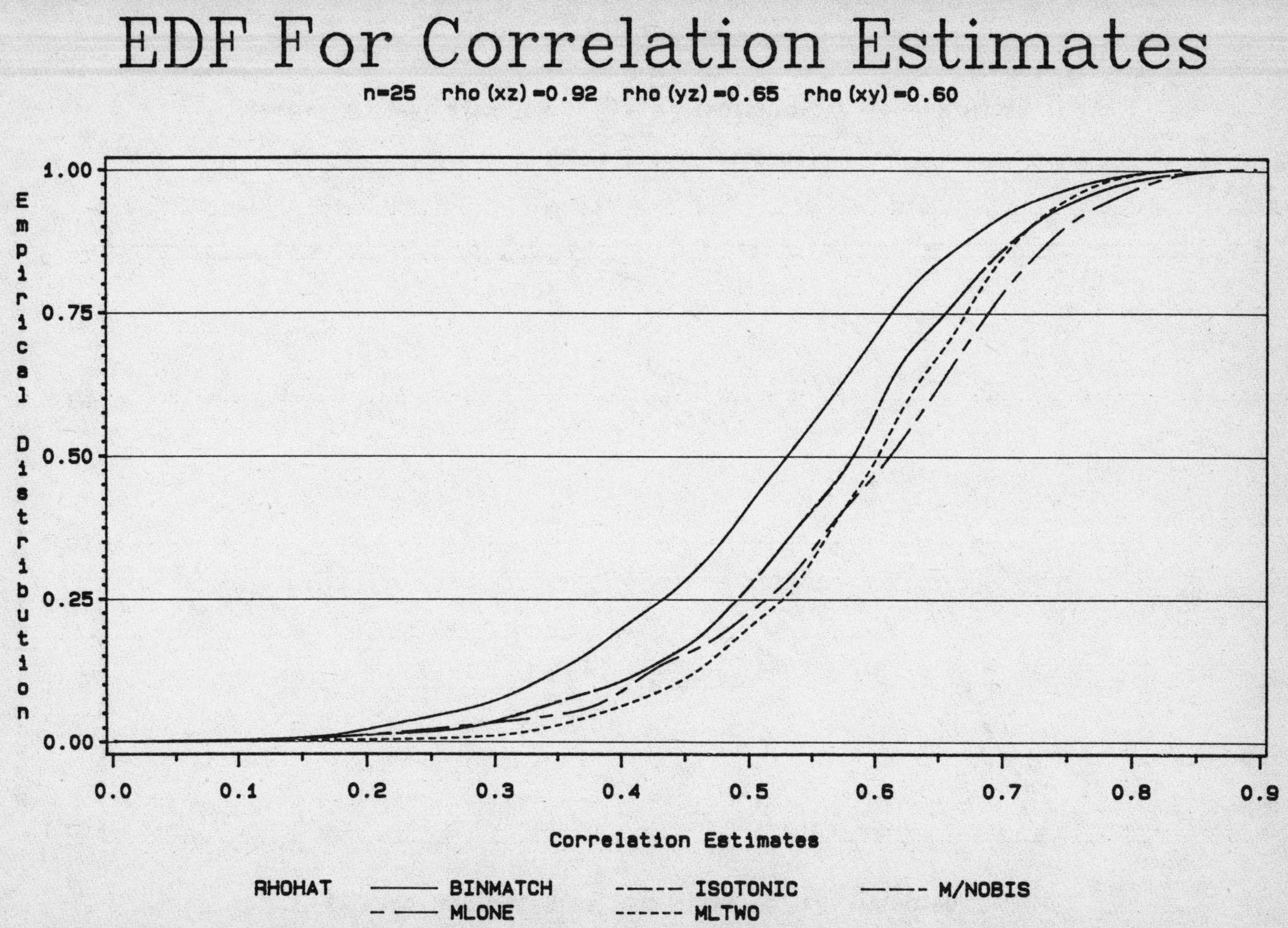
EDF For Correlation Estimates
n=25 rho (xz) =0.92 rho (yz) =0.65 rho (xy) =0.60
Empirical Distribution
1.00
0.75
0.50
0.25
0.00
0.0
0.1
0.2
0.3
0.4
0.5
0.6
0.7
0.8
0.9
Correlation Estimates
RHOHAT
BINMATCH
ISOTONIC
M/NOBIS
MLONE
MLTWO

Figure 3.5

EDF For Correlation Estimates

n=50 rho (xz) =0.92 rho (yz) =0.65 rho (xy) =0.60

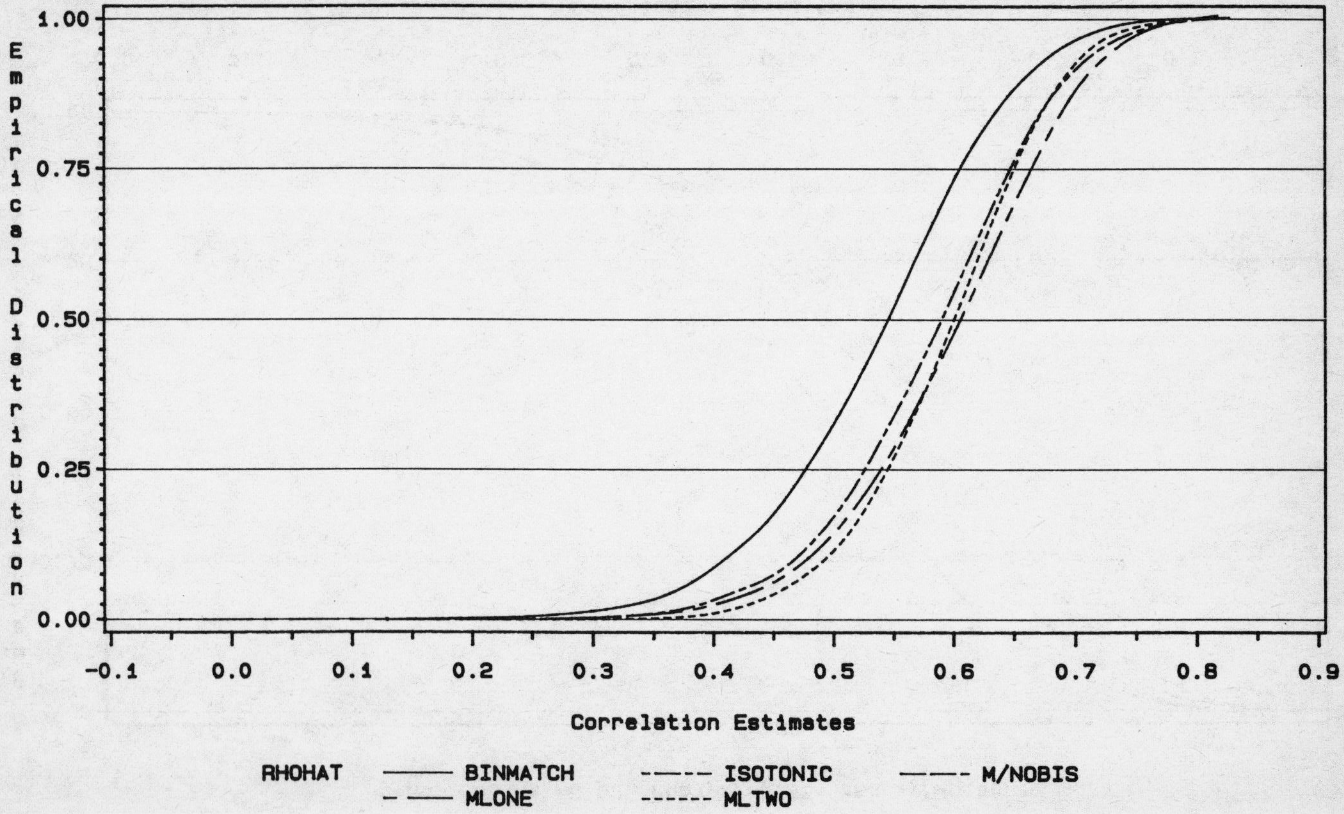

Figure 3.6

EDF For Correlation Estimates

n=10 rho (xz) =0.93 rho (yz) =0.75 rho (xy) =0.70

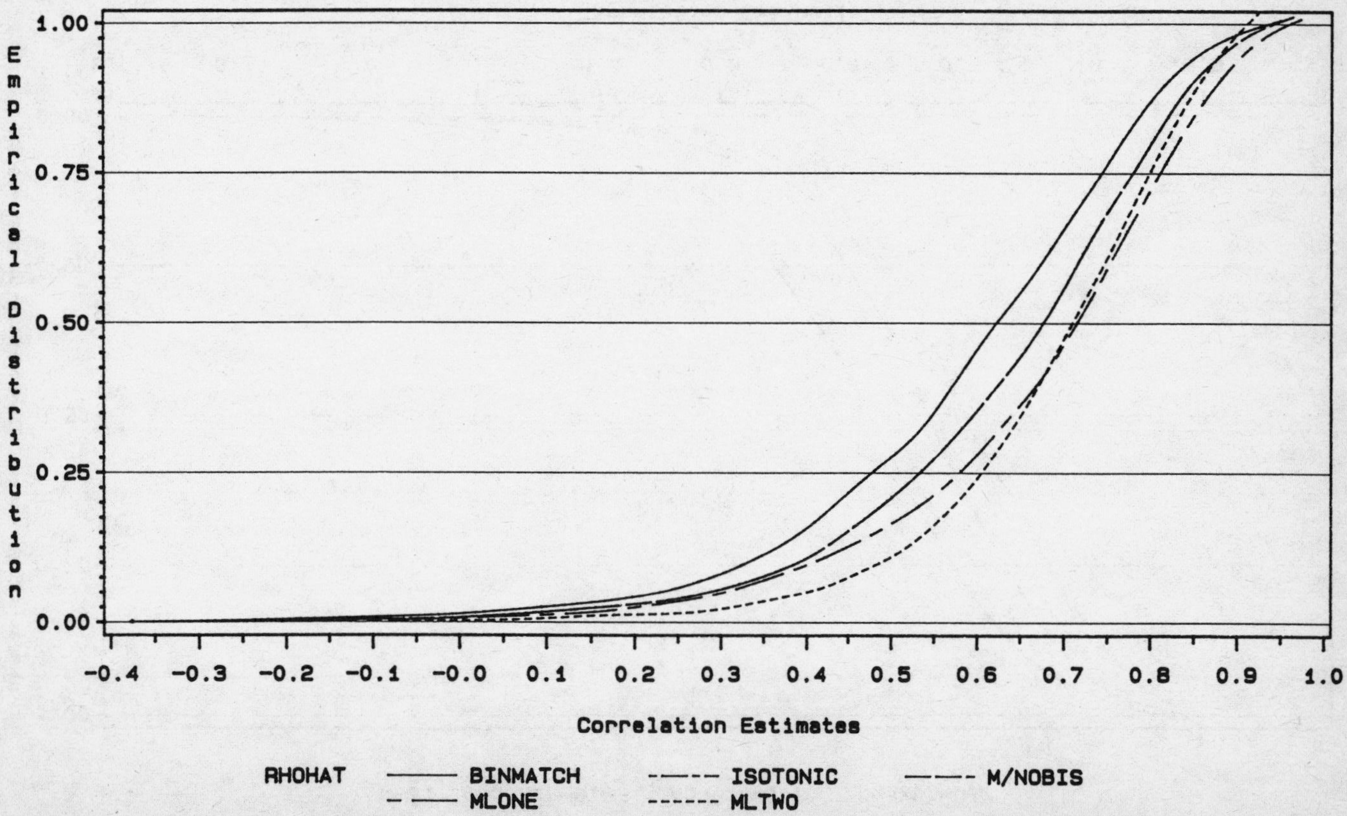

Figure 3.7

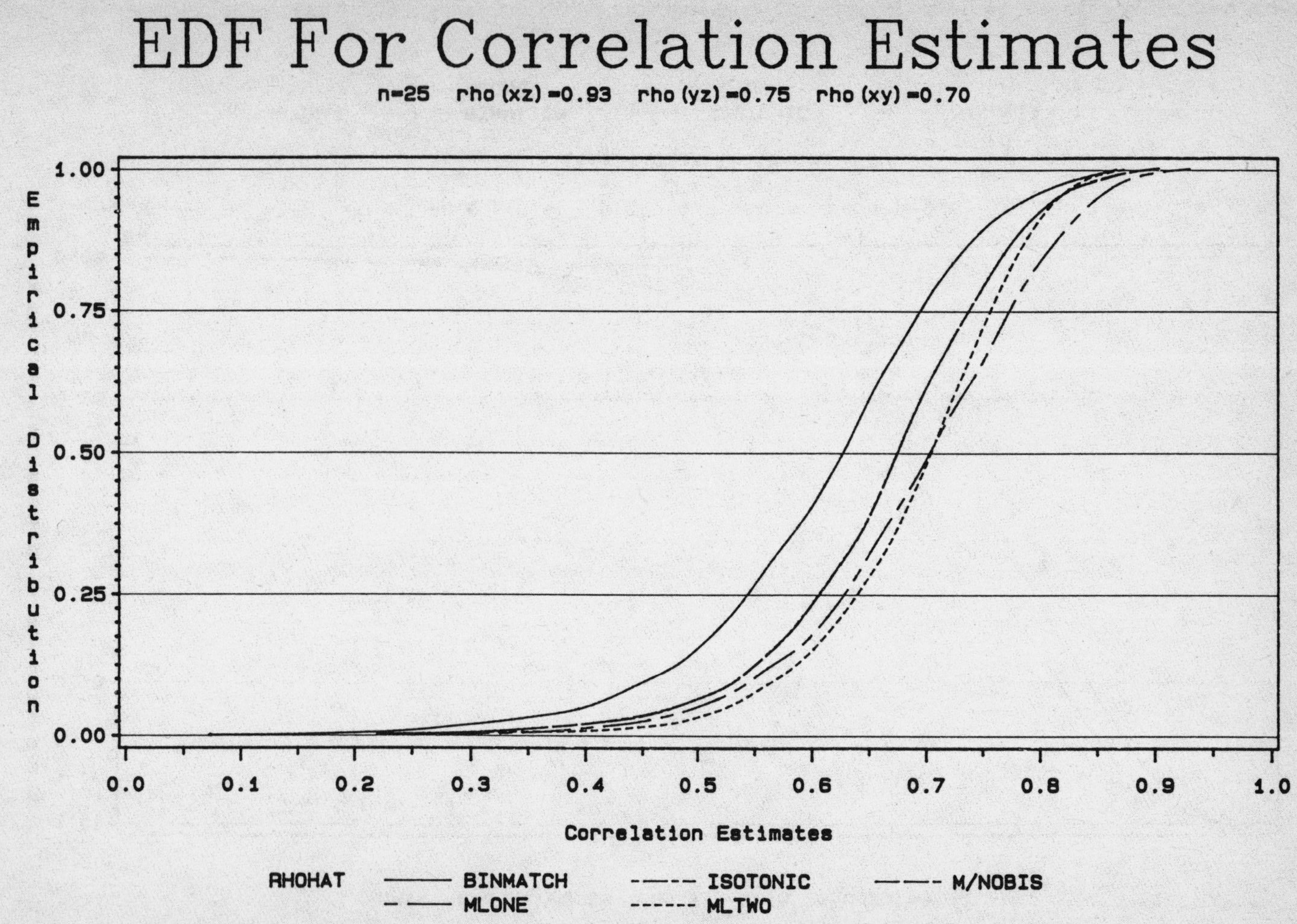
EDF For Correlation Estimates
n=25 rho (xz) =0.93 rho (yz) =0.75 rho (xy) =0.70
Empirical Distribution
1.00
0.75
0.50
0.25
0.00
0.0
0.1
0.2
0.3
0.4
0.5
0.6
0.7
0.8
0.9
1.0
Correlation Estimates
RHOHAT
BINMATCH
ISOTONIC
M/NOBIS
MLONE
MLTWO

Figure 3.8

EDF For Correlation Estimates

n=50 rho (xz) =0.93 rho (yz) =0.75 rho (xy) =0.70

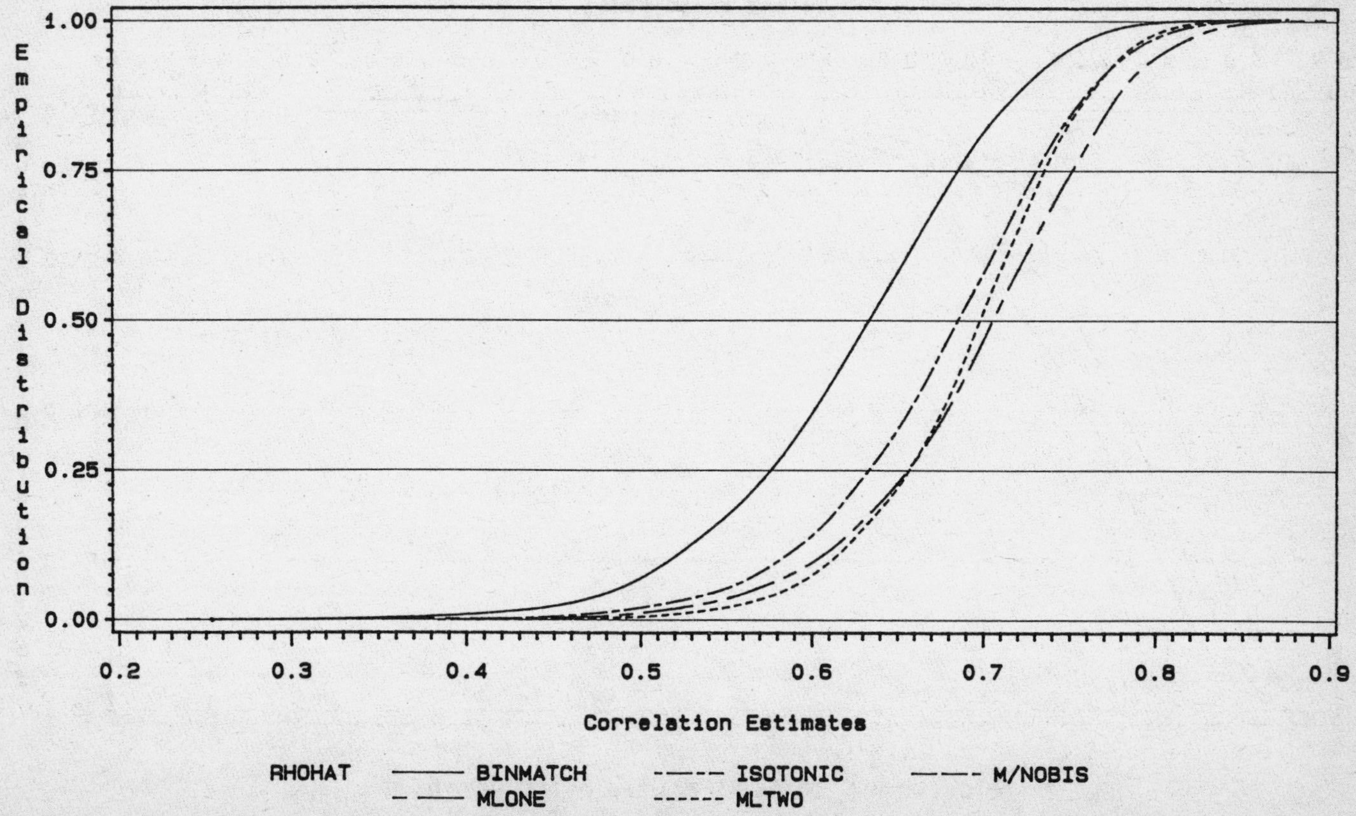

Figure 3.9

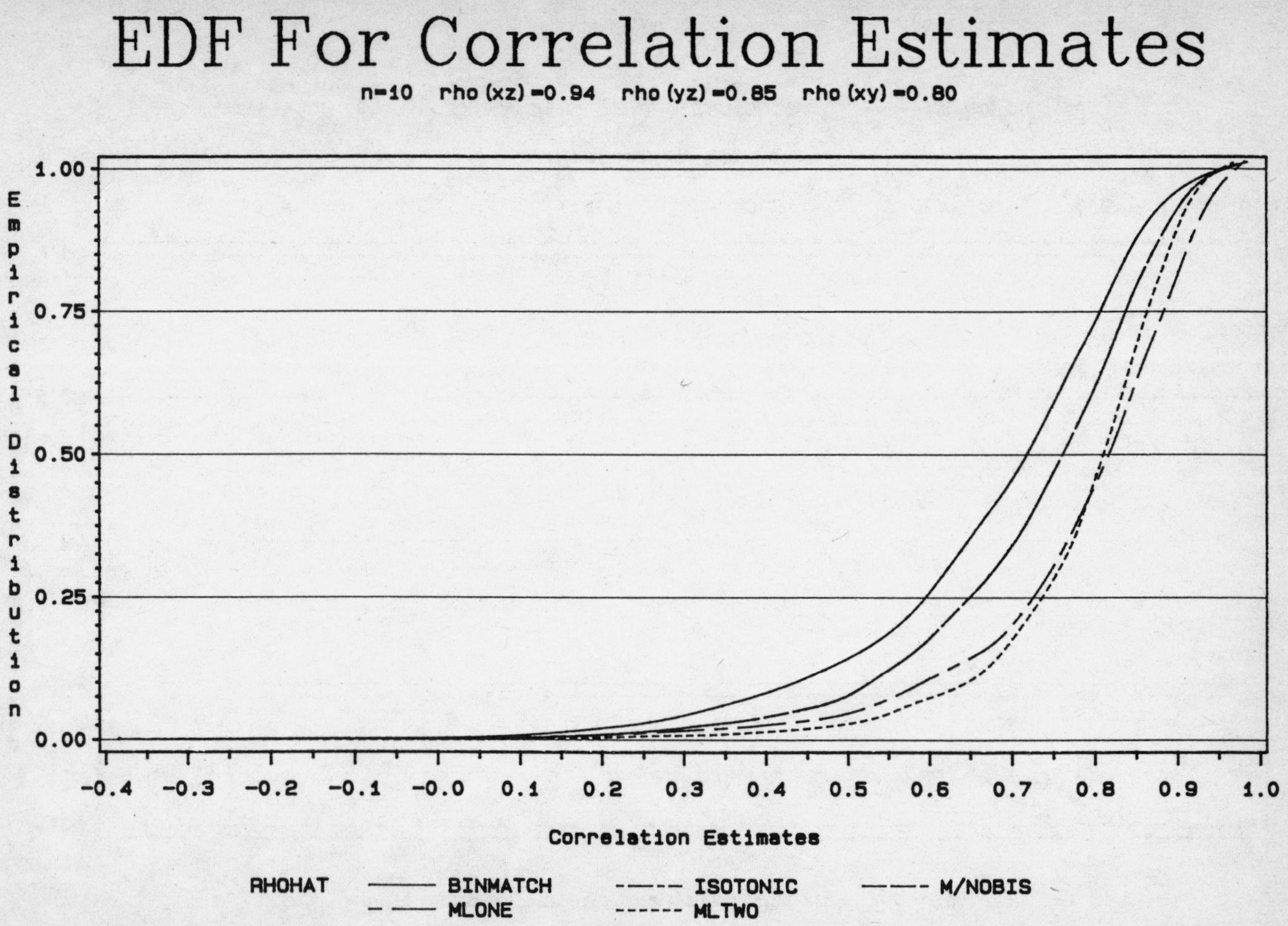
EDF For Correlation Estimates
n=10 rho (xz) =0.94 rho (yz) =0.85 rho (xy) =0.80
Empirical Distribution
1.00
0.75
0.50
0.25
0.00
-0.4 -0.3 -0.2 -0.1 -0.0 0.1 0.2 0.3 0.4 0.5 0.6 0.7 0.8 0.9 1.0
Correlation Estimates
RHOHAT
BINMATCH
MLONE
ISOTONIC
MLTWO
M/NOBIS

Figure 3.10

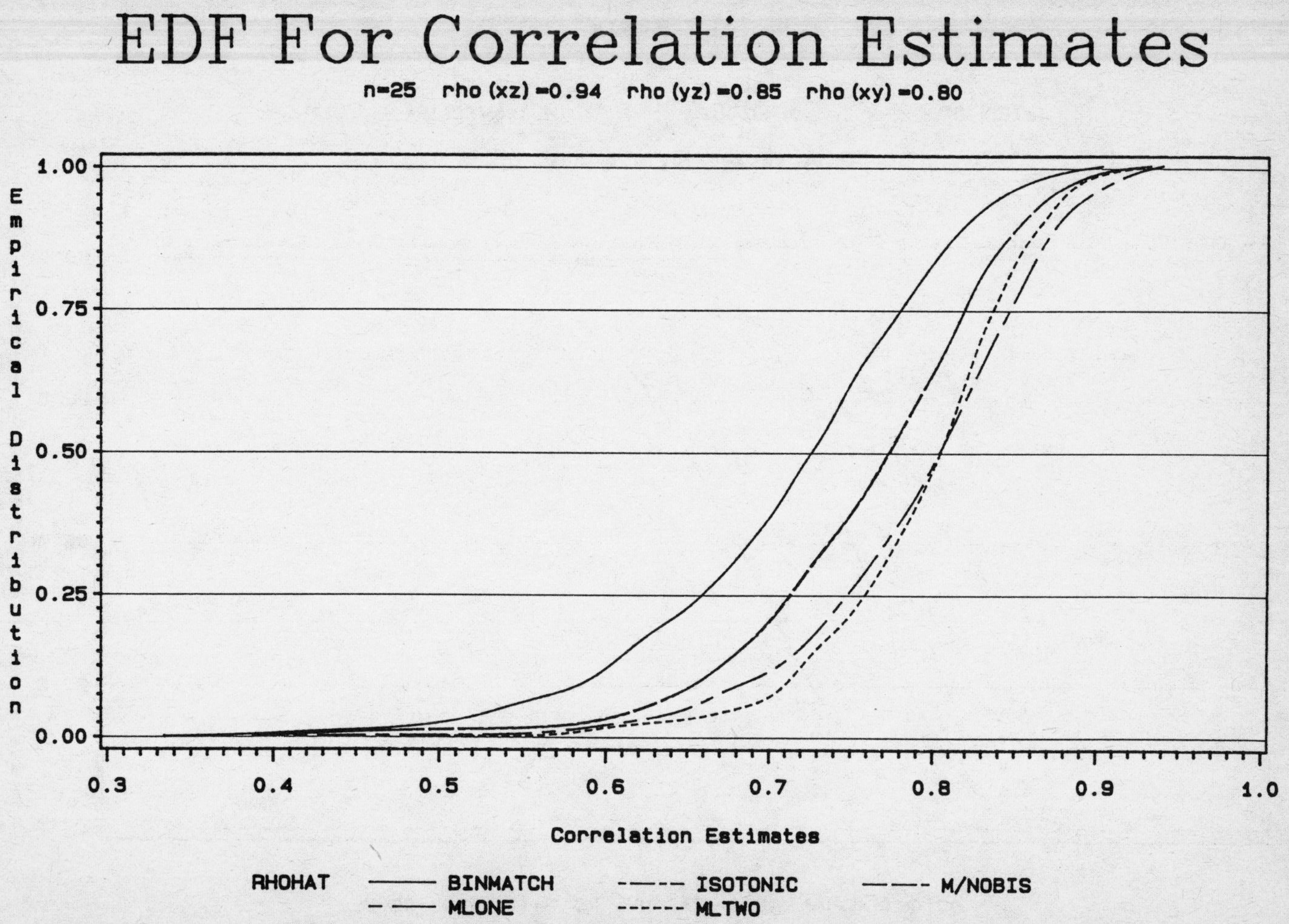
EDF For Correlation Estimates
n=25 rho (xz) =0.94 rho (yz) =0.85 rho (xy) =0.80
Empirical Distribution
1.00
0.75
0.50
0.25
0.00
0.3
0.4
0.5
0.6
0.7
0.8
0.9
1.0
Correlation Estimates
RHOHAT
BINMATCH
ISOTONIC
M/NOBIS
MLONE
MLTWO

Figure 3.11

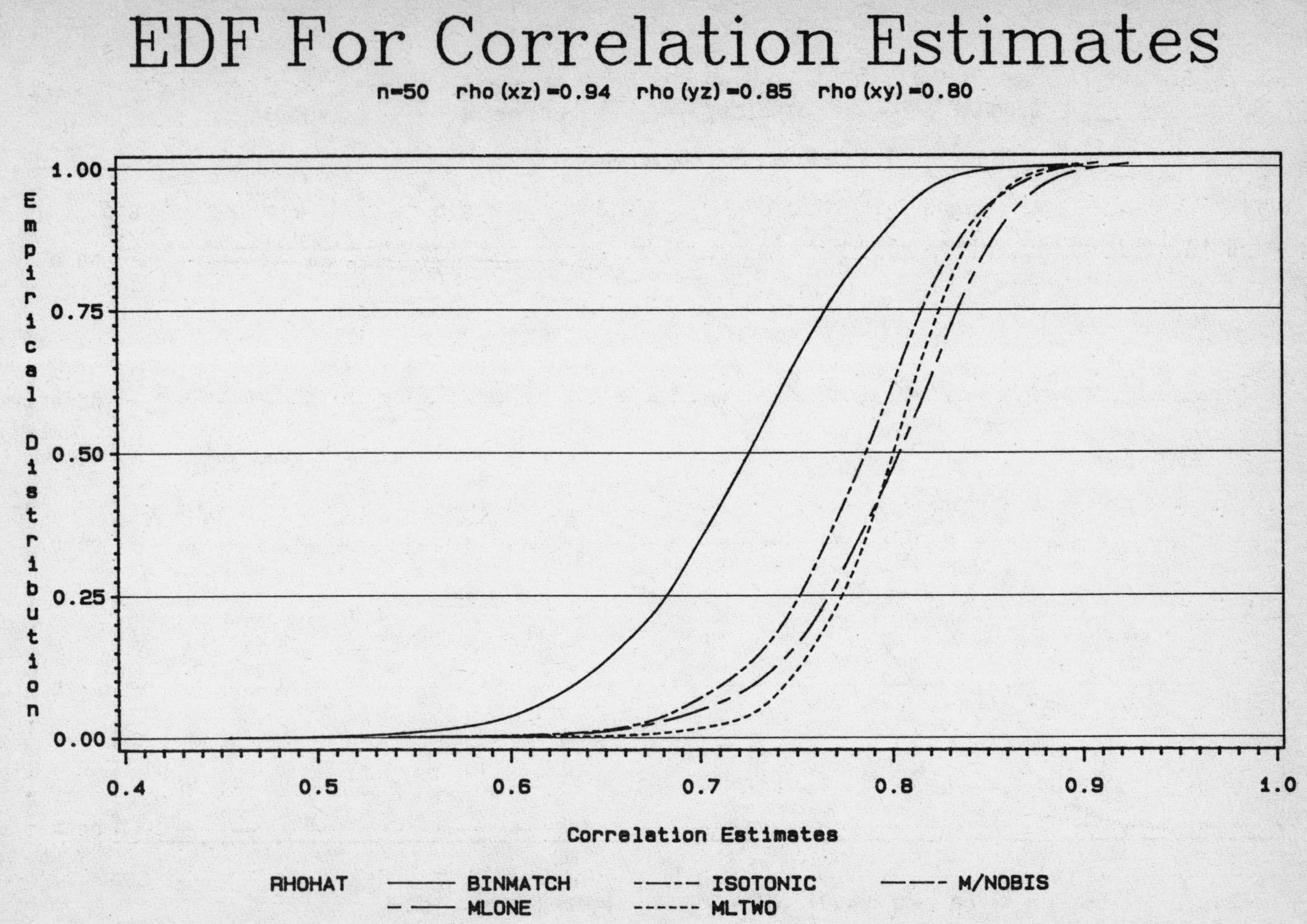
EDF For Correlation Estimates
n=50 rho(xz)=0.94 rho(yz)=0.85 rho(xy)=0.80
Empirical Distribution
1.00
0.75
0.50
0.25
0.00
0.4
0.5
0.6
0.7
0.8
0.9
1.0
Correlation Estimates
RHOHAT
BINMATCH
ISOTONIC
M/NOBIS
MLONE
MLTWO

Figure 3.12

EDF For Correlation Estimates

n=10 rho (xz) =0.95 rho (yz) =0.95 rho (xy) =0.90

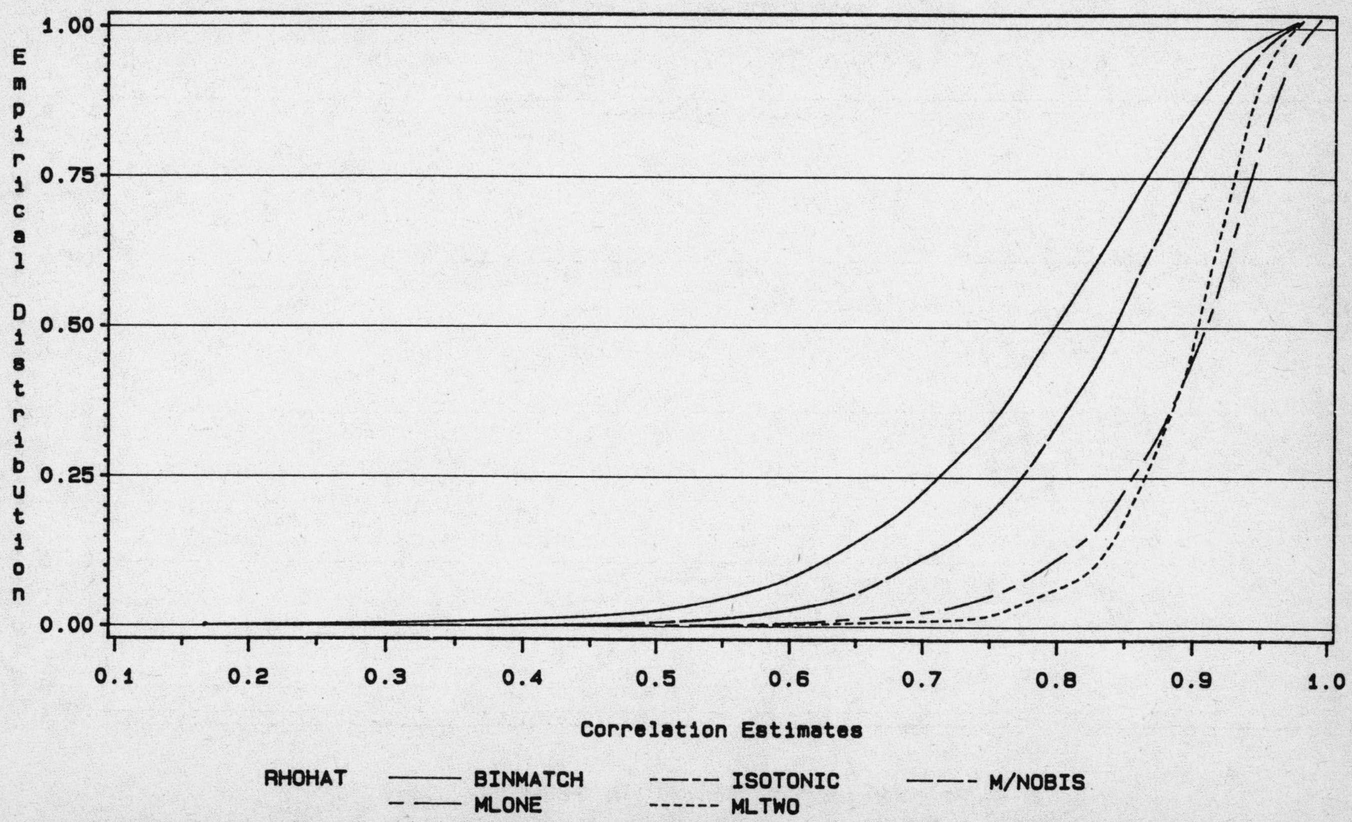

Figure 3.13

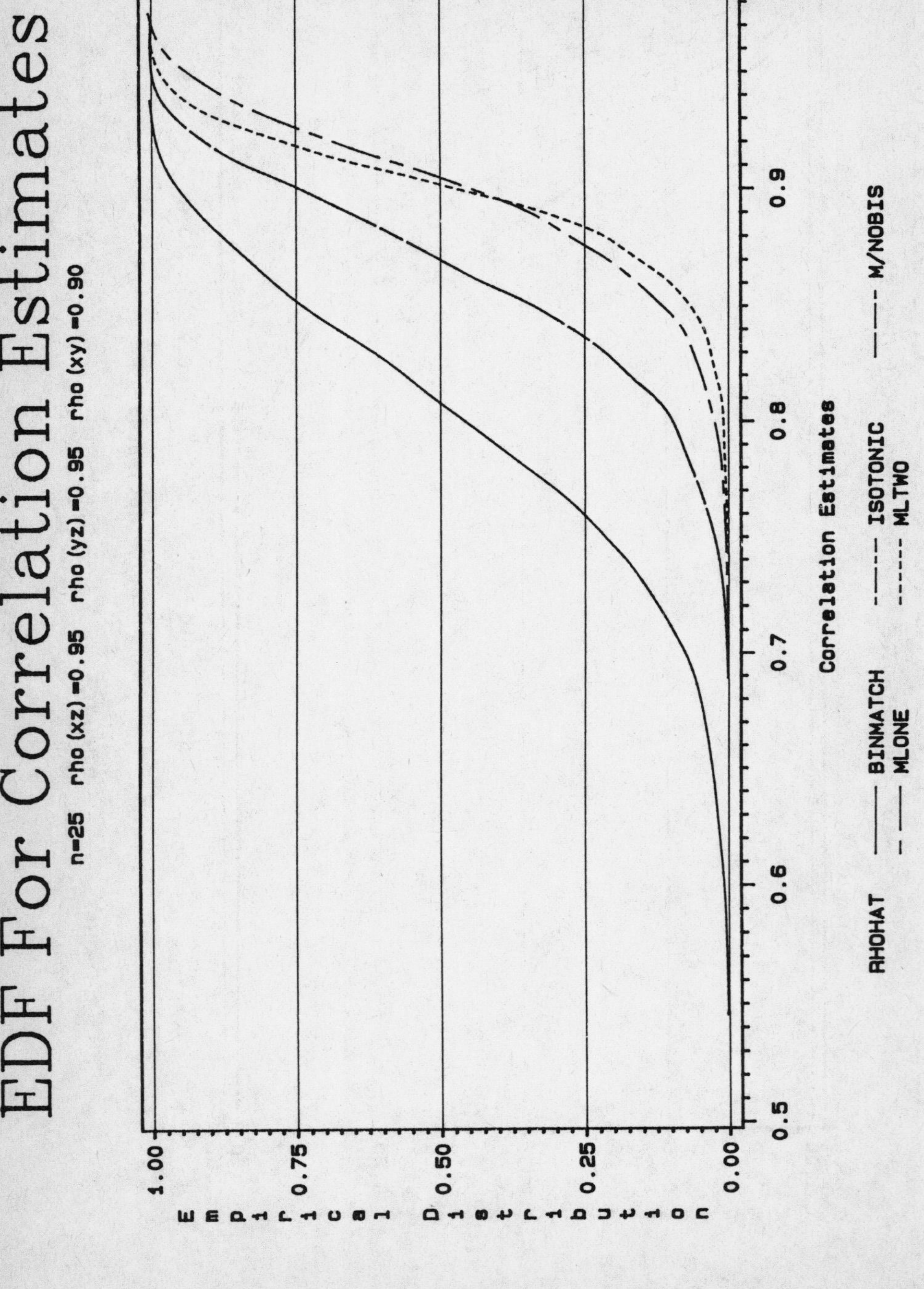
EDF For Correlation Estimates
n=25 rho(xz)=0.95 rho(yz)=0.95 rho(xy)=0.90
Empirical Distribution
1.00
0.75
0.50
0.25
0.00
0.5
0.6
0.7
0.8
0.9
1.0
Correlation Estimates
RHOHAT
BINMATCH
ISOTONIC
M/NOBIS
MLONE
MLTWO

Figure 3.14

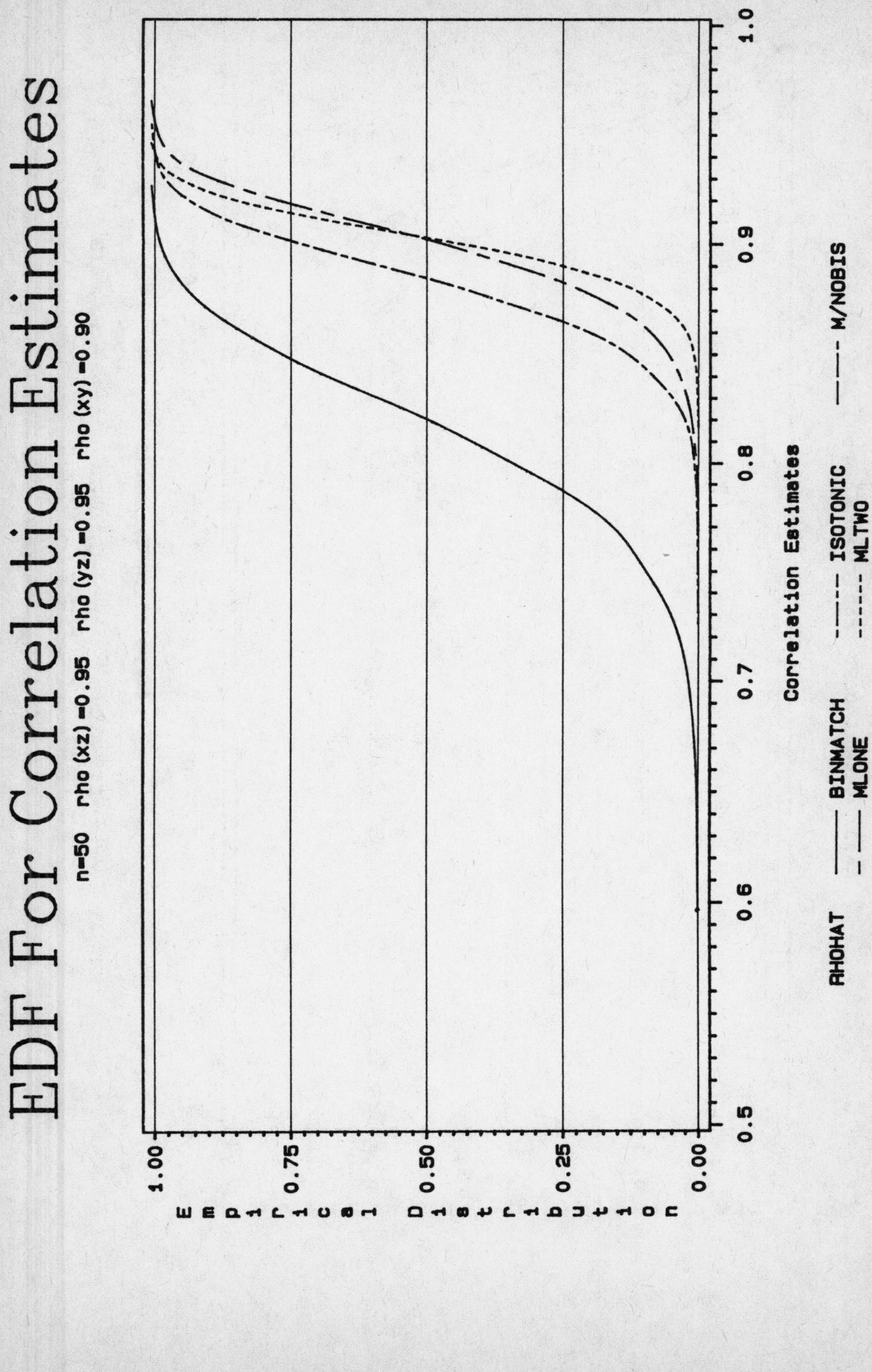

EDF For Correlation Estimates
n=50 rho(xz)=0.95 rho(yz)=0.95 rho(xy)=0.90
Empirical Distribution
1.00
0.75
0.50
0.25
0.00
0.5
0.6
0.7
0.8
0.9
1.0
Correlation Estimates
RHOHAT
BINMATCH
ISOTONIC
M/NOBIS
MLONE
MLTWO

Figure 3.15

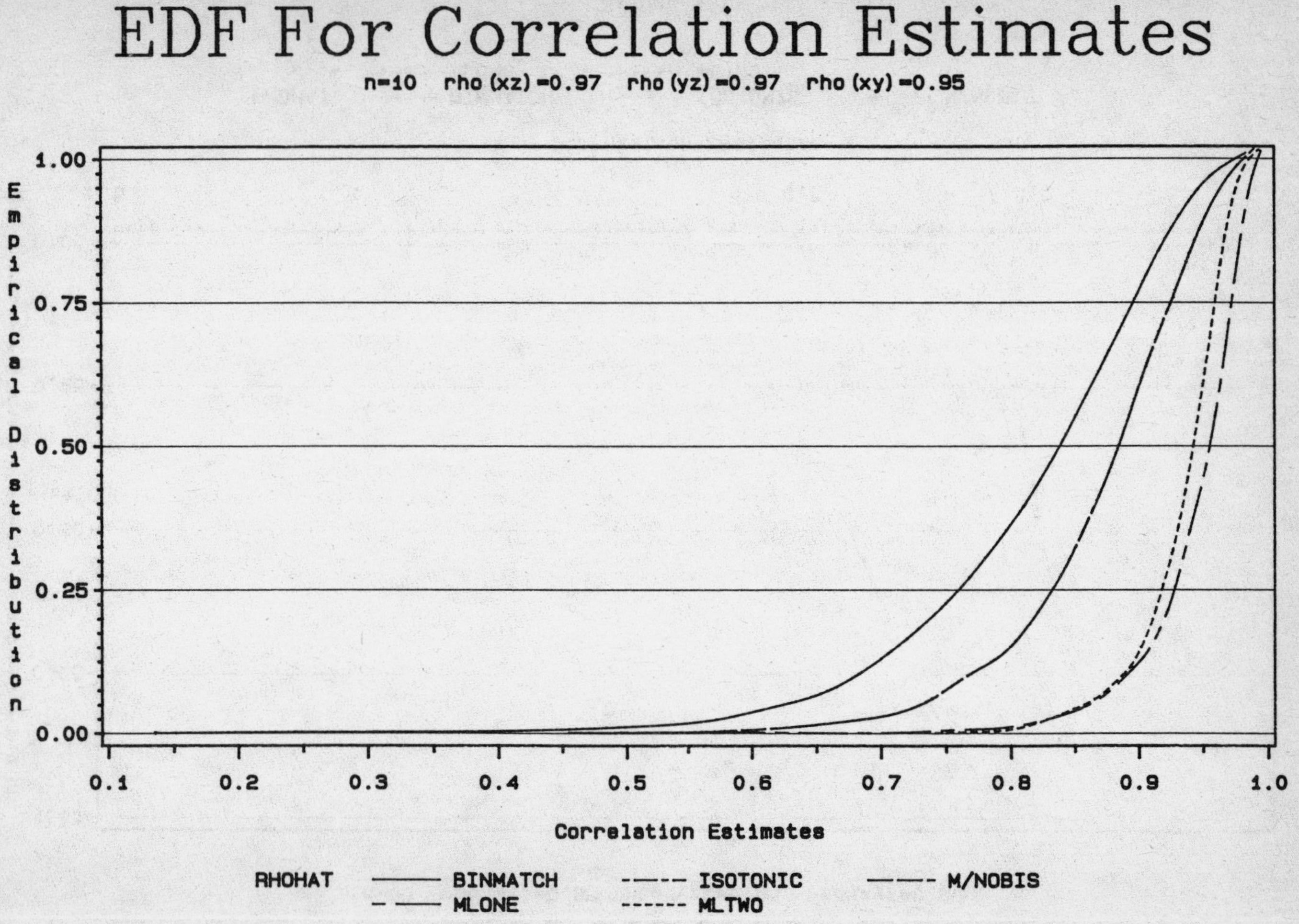
EDF For Correlation Estimates
n=10 rho (xz) =0.97 rho (yz) =0.97 rho (xy) =0.95
Empirical Distribution
1.00
0.75
0.50
0.25
0.00
0.1
0.2
0.3
0.4
0.5
0.6
0.7
0.8
0.9
1.0
Correlation Estimates
RHOHAT
BINMATCH
ISOTONIC
M/NOBIS
MLONE
MLTWO

Figure 3.16

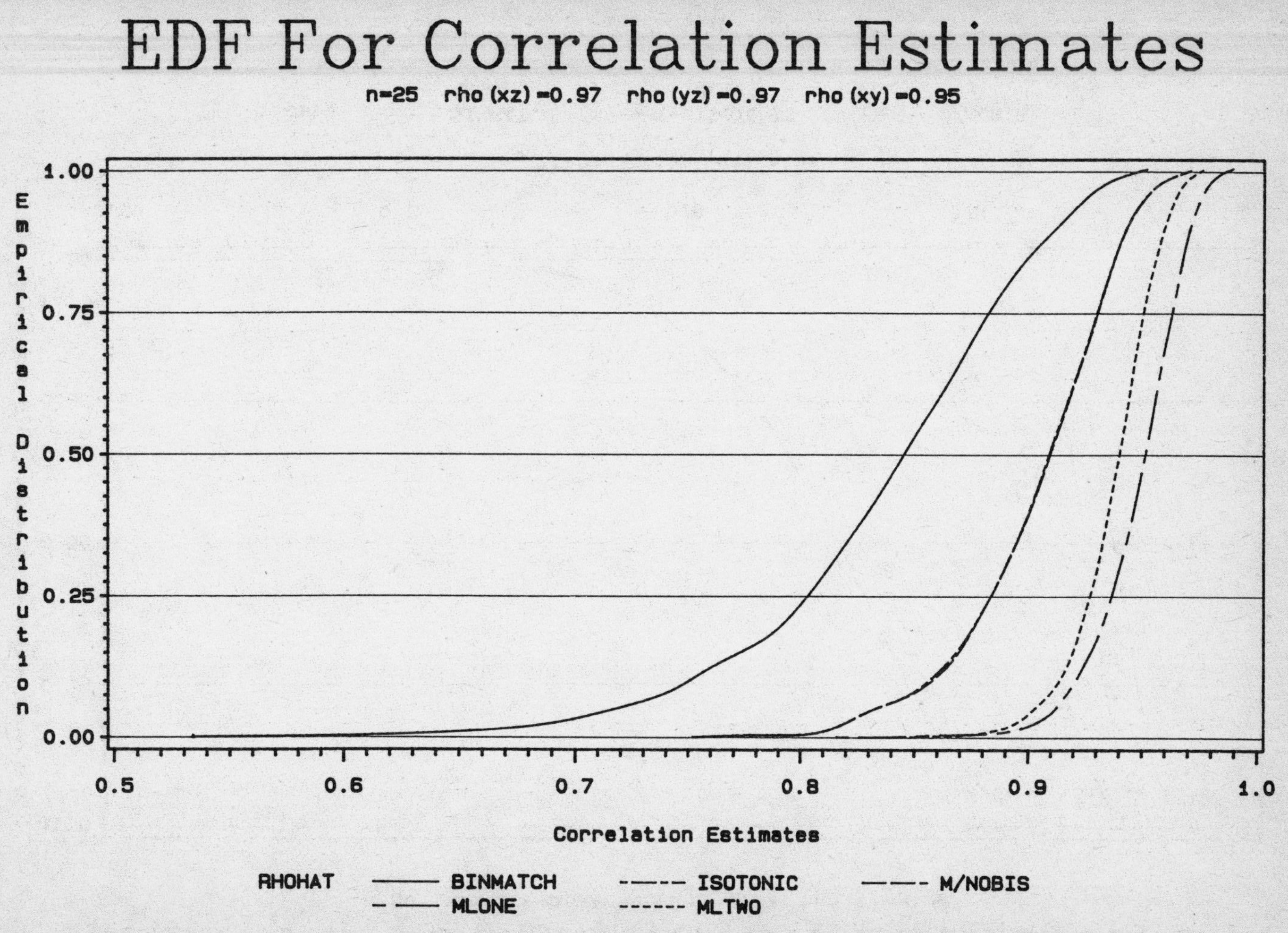
EDF For Correlation Estimates
n=25 rho (xz) =0.97 rho (yz) =0.97 rho (xy) =0.95
Empirical Distribution
1.00
0.75
0.50
0.25
0.00
0.5
0.6
0.7
0.8
0.9
1.0
Correlation Estimates
RHOHAT
BINMATCH
ISOTONIC
M/NOBIS
MLONE
MLTWO

Figure 3.17

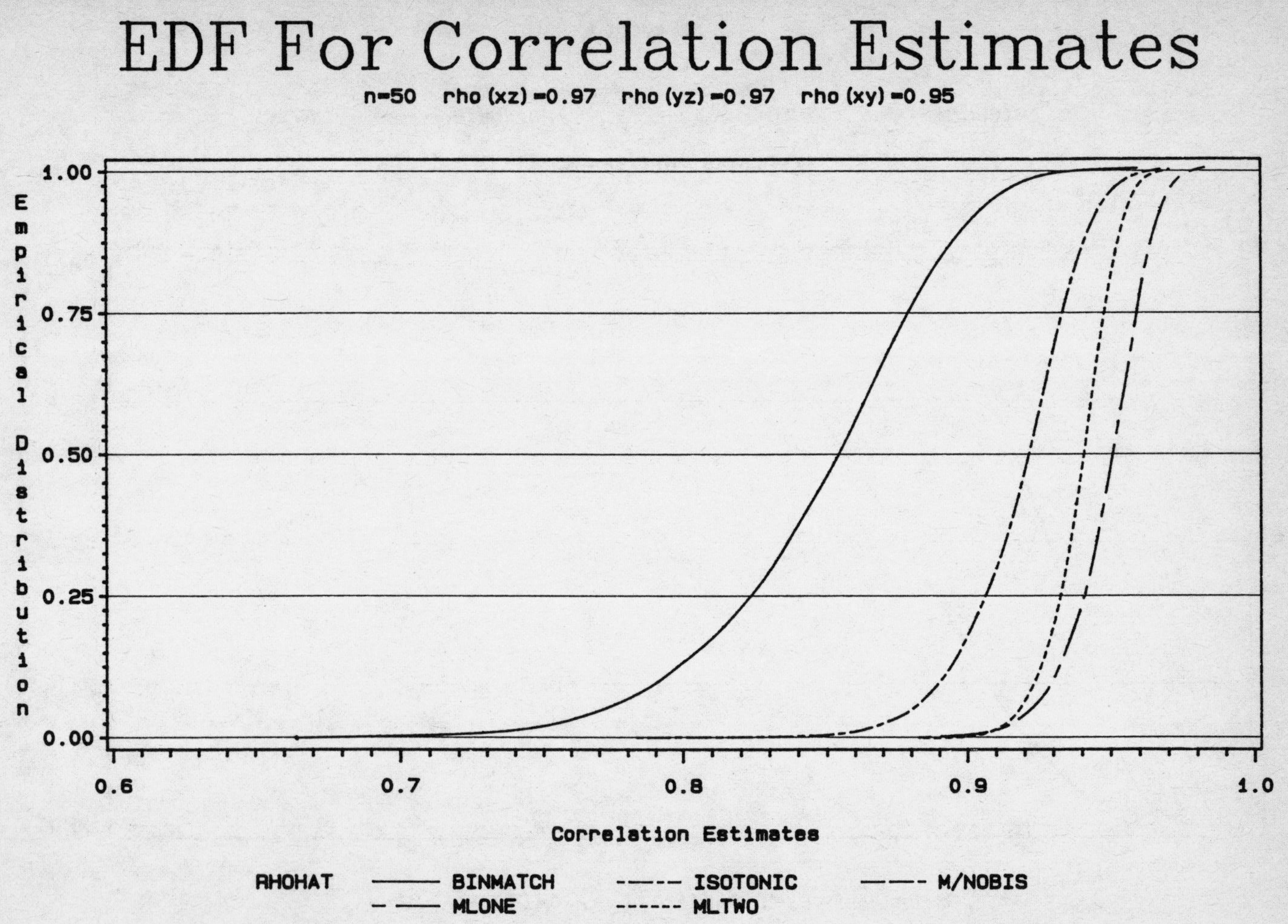

EDF For Correlation Estimates
n=50 rho (xz) =0.97 rho (yz) =0.97 rho (xy) =0.95
Empirical Distribution
1.00
0.75
0.50
0.25
0.00
0.6
0.7
0.8
0.9
1.0
Correlation Estimates
RHOHAT
BINMATCH
ISOTONIC
M/NOBIS
MLONE
MLTWO

Figure 3.18

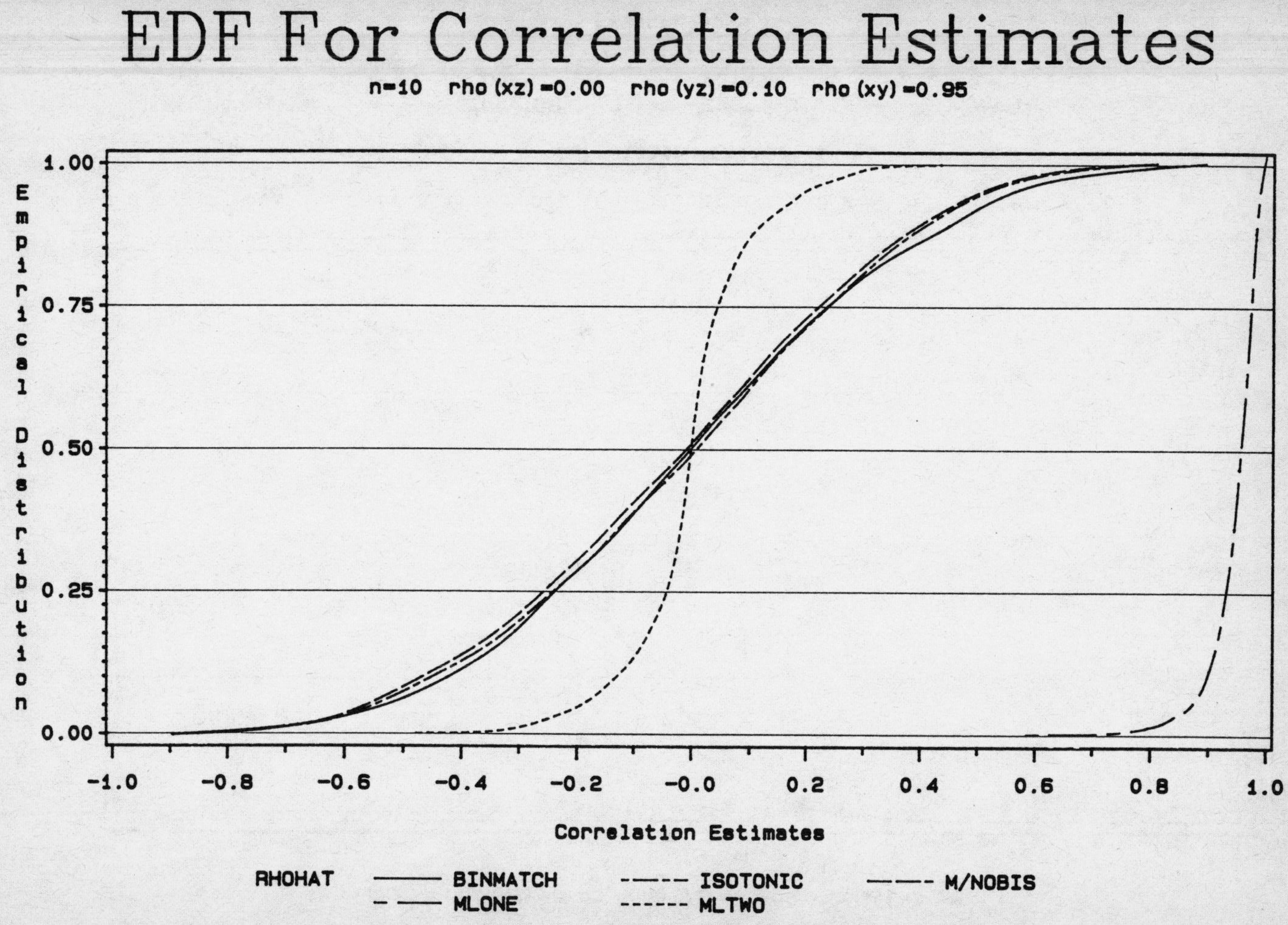
EDF For Correlation Estimates
n=10 rho (xz) =0.00 rho (yz) =0.10 rho (xy) =0.95
Empirical Distribution
1.00
0.75
0.50
0.25
0.00
-1.0
-0.8
-0.6
-0.4
-0.2
-0.0
0.2
0.4
0.6
0.8
1.0
Correlation Estimates
RHOHAT
BINMATCH
ISOTONIC
M/NOBIS
MLONE
MLTWO

Figure 3.19

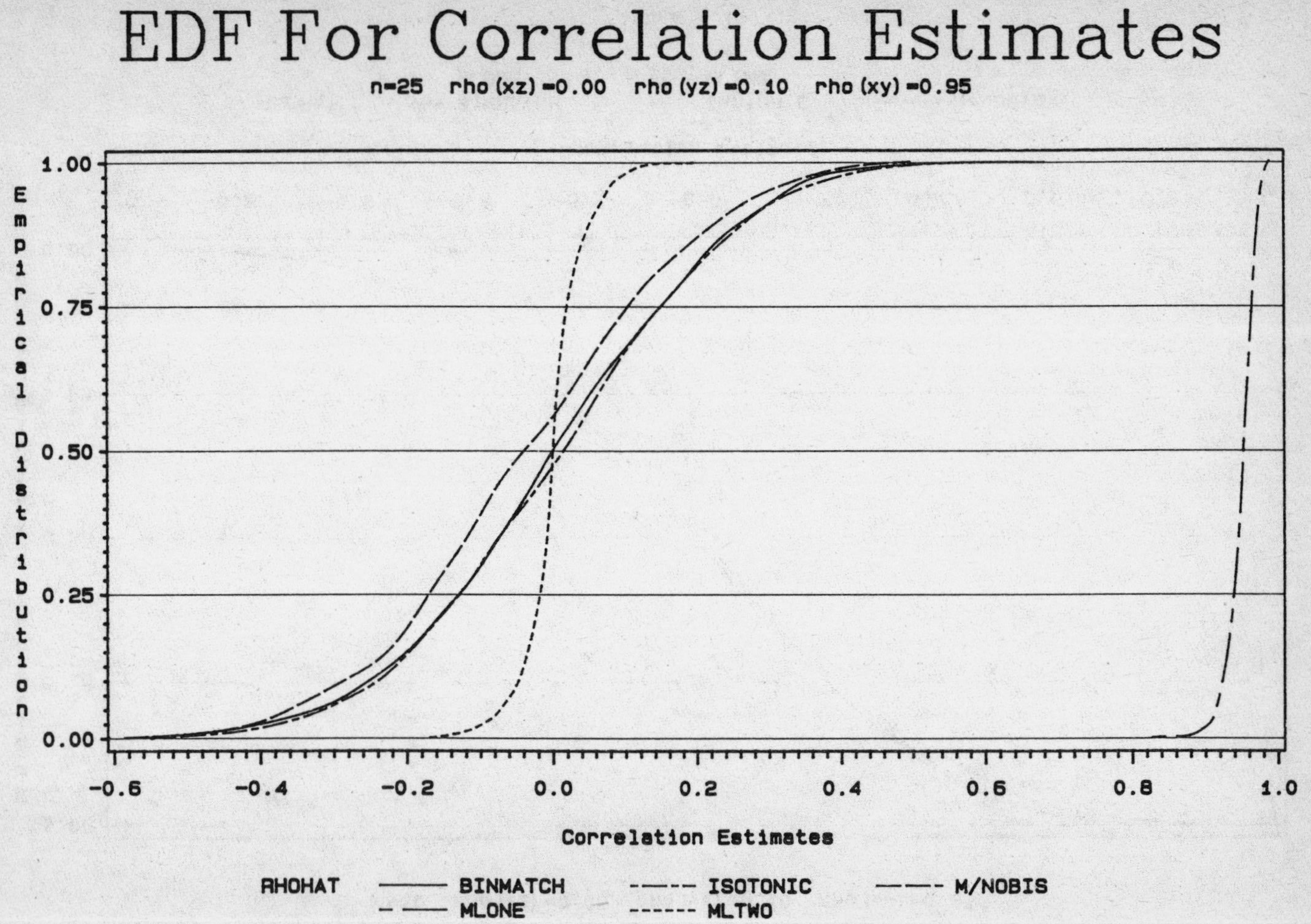
EDF For Correlation Estimates
n=25 rho (xz) =0.00 rho (yz) =0.10 rho (xy) =0.95
Empirical Distribution
1.00
0.75
0.50
0.25
0.00
-0.6
-0.4
-0.2
0.0
0.2
0.4
0.6
0.8
1.0
Correlation Estimates
RHOHAT
BINMATCH
ISOTONIC
M/NOBIS
MLONE
MLTWO

Figure 3.20

EDF For Correlation Estimates

n=50 rho (xz) =0.00 rho (yz) =0.10 rho (xy) =0.95

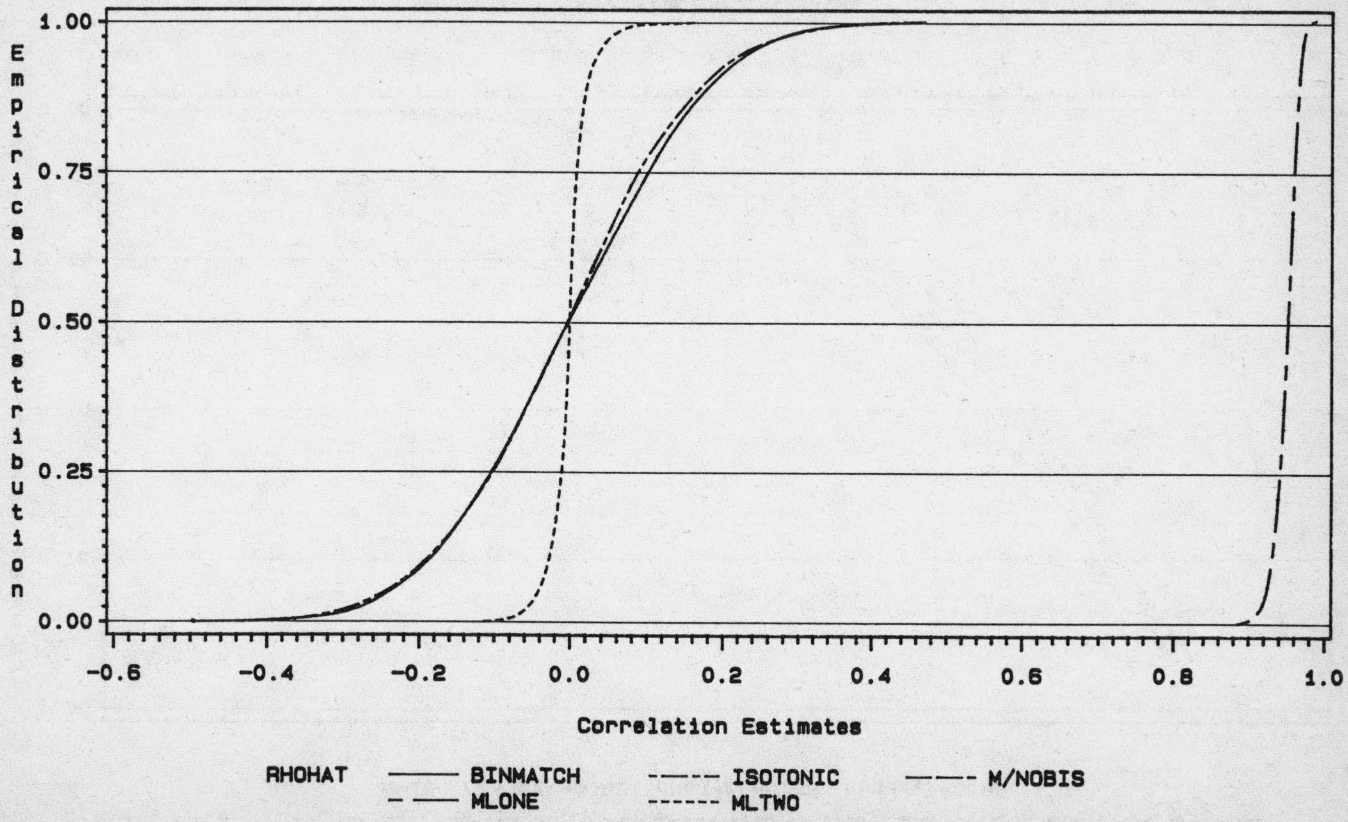

Figure 3.21

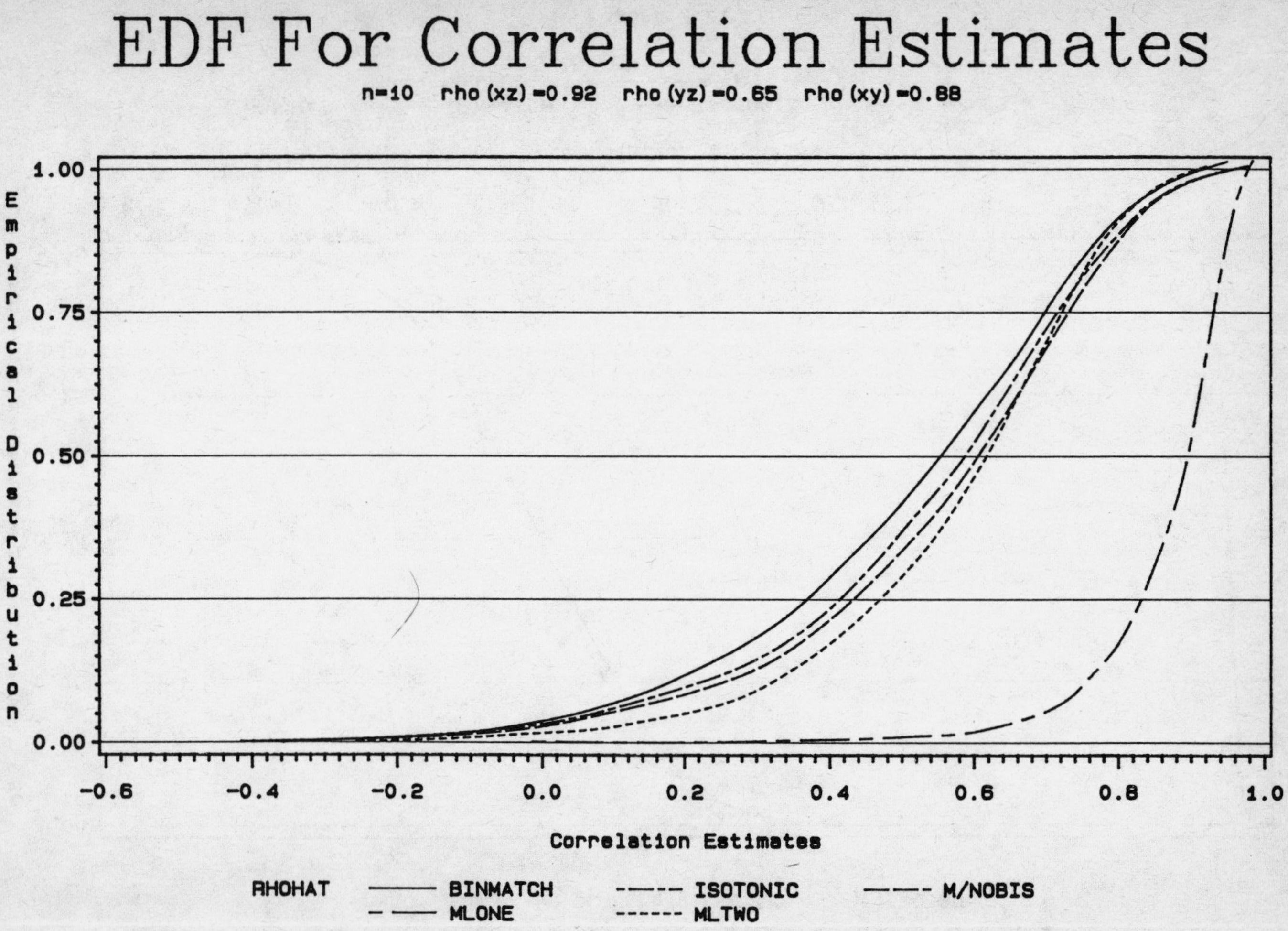
EDF For Correlation Estimates
n=10 rho (xz) =0.92 rho (yz) =0.65 rho (xy) =0.88
Empirical Distribution
1.00
0.75
0.50
0.25
0.00
-0.6
-0.4
-0.2
0.0
0.2
0.4
0.6
0.8
1.0
Correlation Estimates
RHOHAT
BINMATCH
ISOTONIC
M/NOBIS
MLONE
MLTWO

Figure 3.22

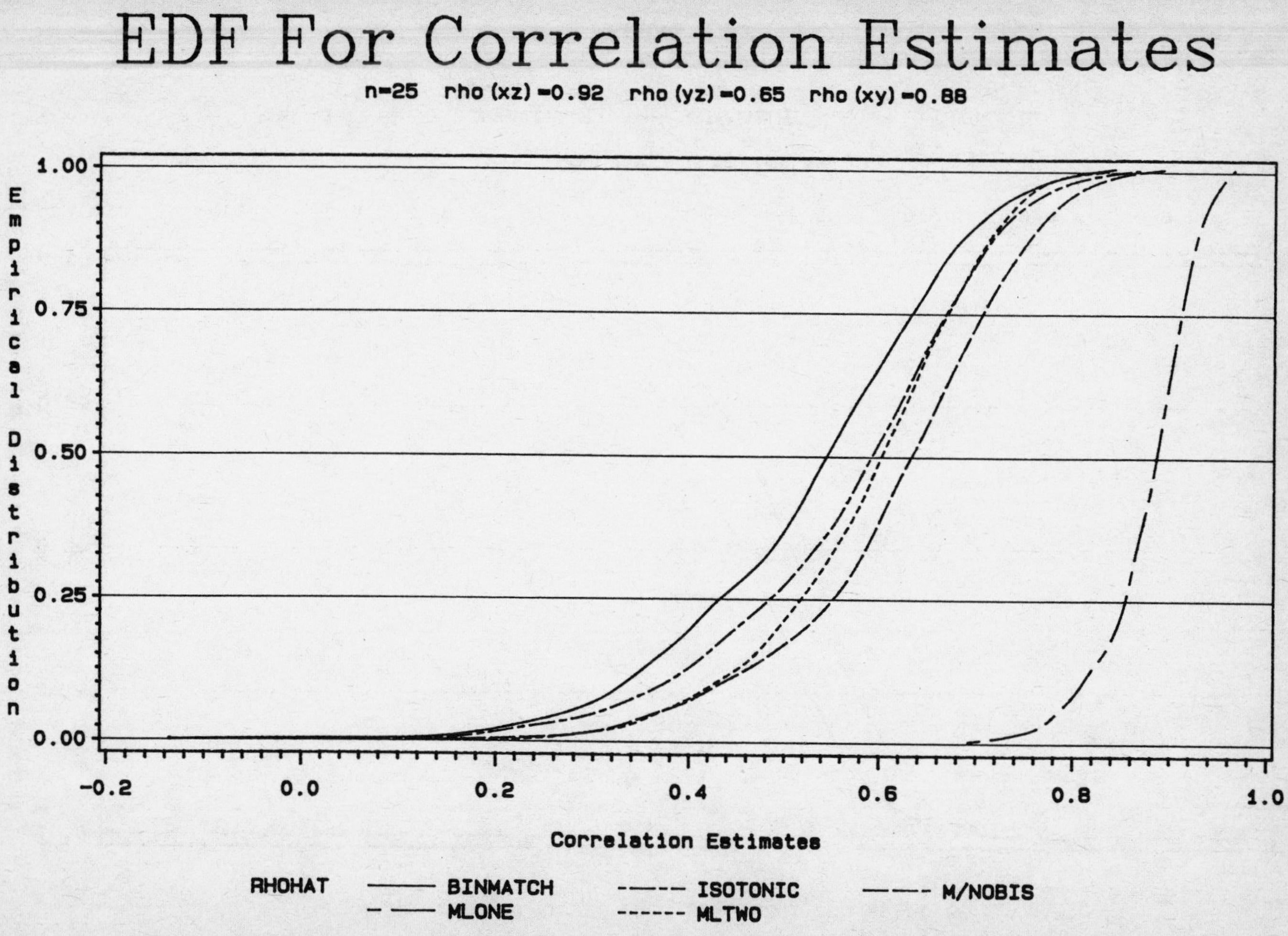
EDF For Correlation Estimates
n=25 rho (xz) =0.92 rho (yz) =0.65 rho (xy) =0.88
Empirical Distribution
1.00
0.75
0.50
0.25
0.00
-0.2
0.0
0.2
0.4
0.6
0.8
1.0
Correlation Estimates
RHOHAT
BINMATCH
ISOTONIC
M/NOBIS
MLONE
MLTWO

Figure 3.23

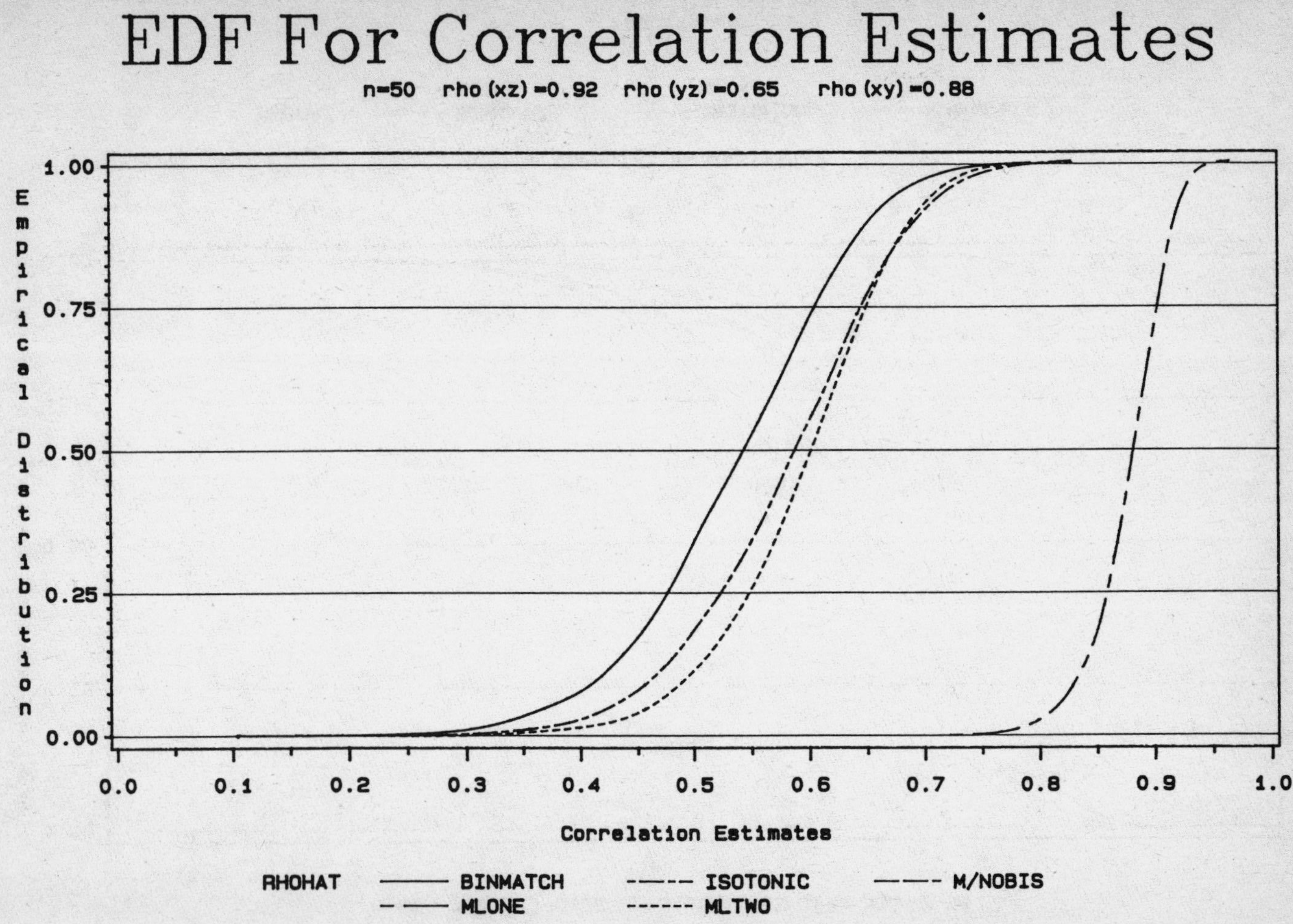
EDF For Correlation Estimates
n=50 rho (xz) =0.92 rho (yz) =0.65 rho (xy) =0.88
Empirical Distribution
1.00
0.75
0.50
0.25
0.00
0.0
0.1
0.2
0.3
0.4
0.5
0.6
0.7
0.8
0.9
1.0
Correlation Estimates
RHOHAT
BINMATCH
ISOTONIC
M/NOBIS
MLONE
MLTWO

Figure 3.24

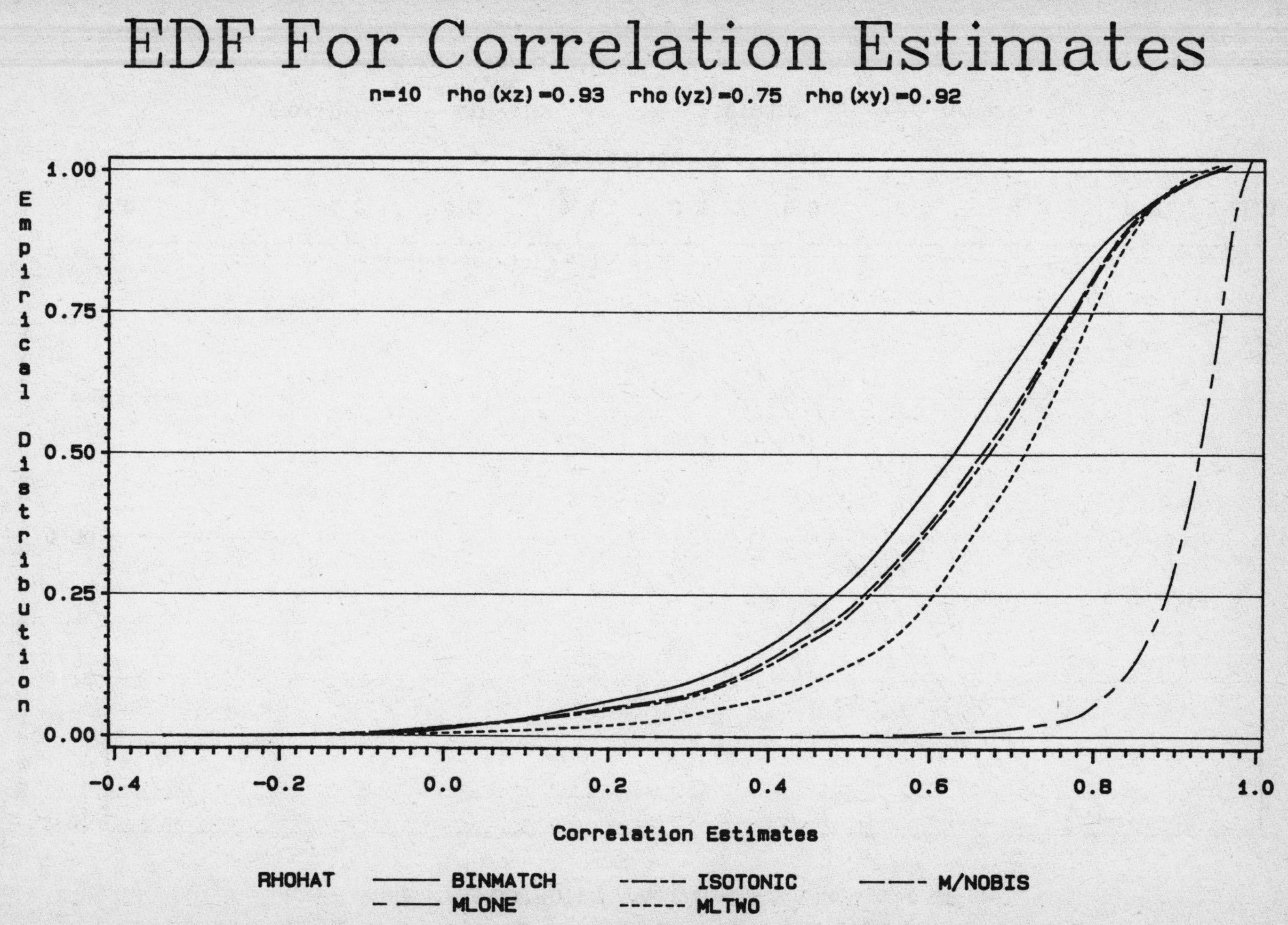
EDF For Correlation Estimates
n=10 rho (xz) =0.93 rho (yz) =0.75 rho (xy) =0.92
Empirical Distribution
1.00
0.75
0.50
0.25
0.00
-0.4
-0.2
0.0
0.2
0.4
0.6
0.8
1.0
Correlation Estimates
RHOHAT
BINMATCH
ISOTONIC
M/NOBIS
MLONE
MLTWO

Figure 3.25

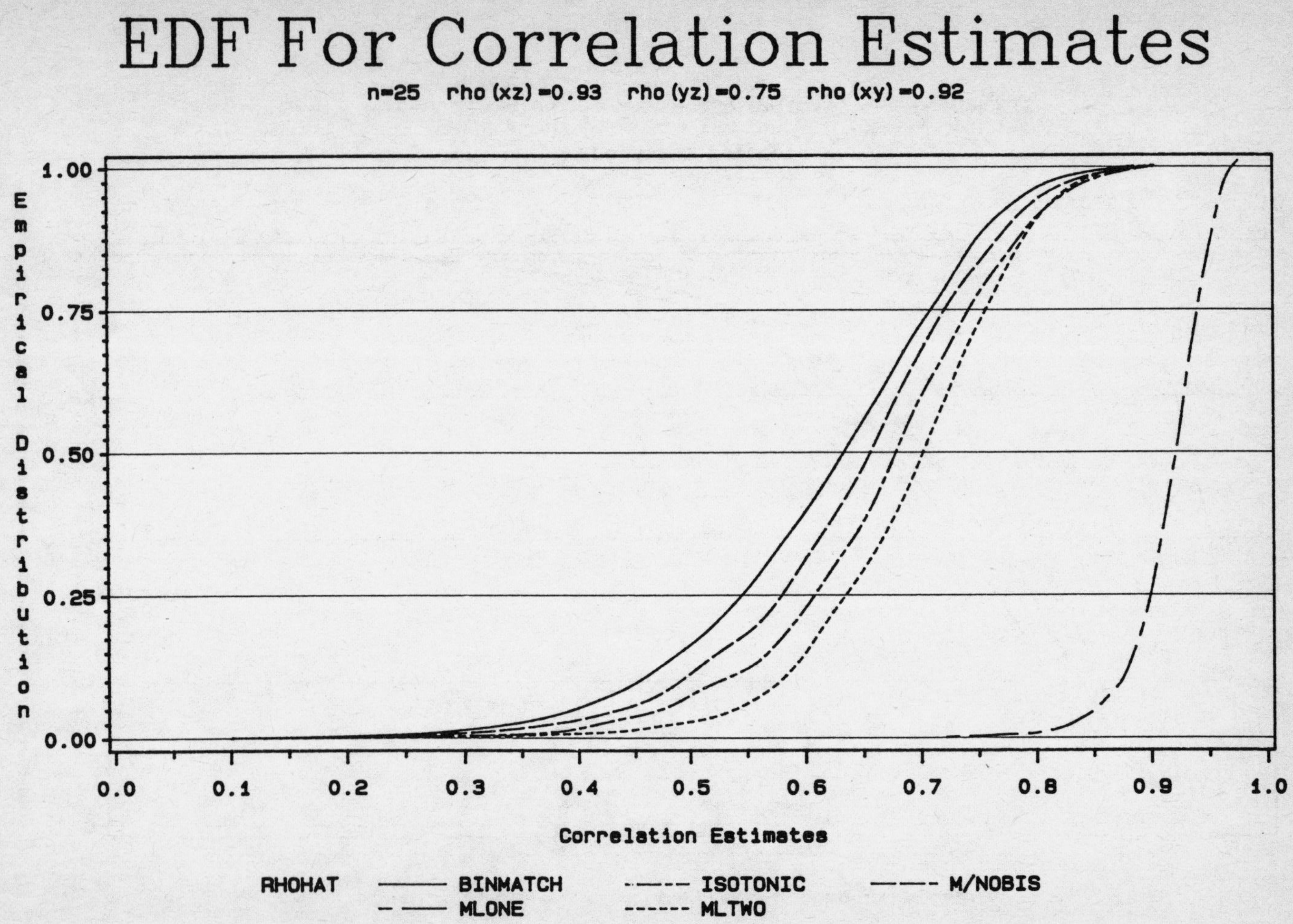
EDF For Correlation Estimates
n=25 rho(xz)=0.93 rho(yz)=0.75 rho(xy)=0.92
Empirical Distribution
1.00
0.75
0.50
0.25
0.00
0.0
0.1
0.2
0.3
0.4
0.5
0.6
0.7
0.8
0.9
1.0
Correlation Estimates
RHOHAT
BINMATCH
ISOTONIC
M/NOBIS
MLONE
MLTWO

Figure 3.26

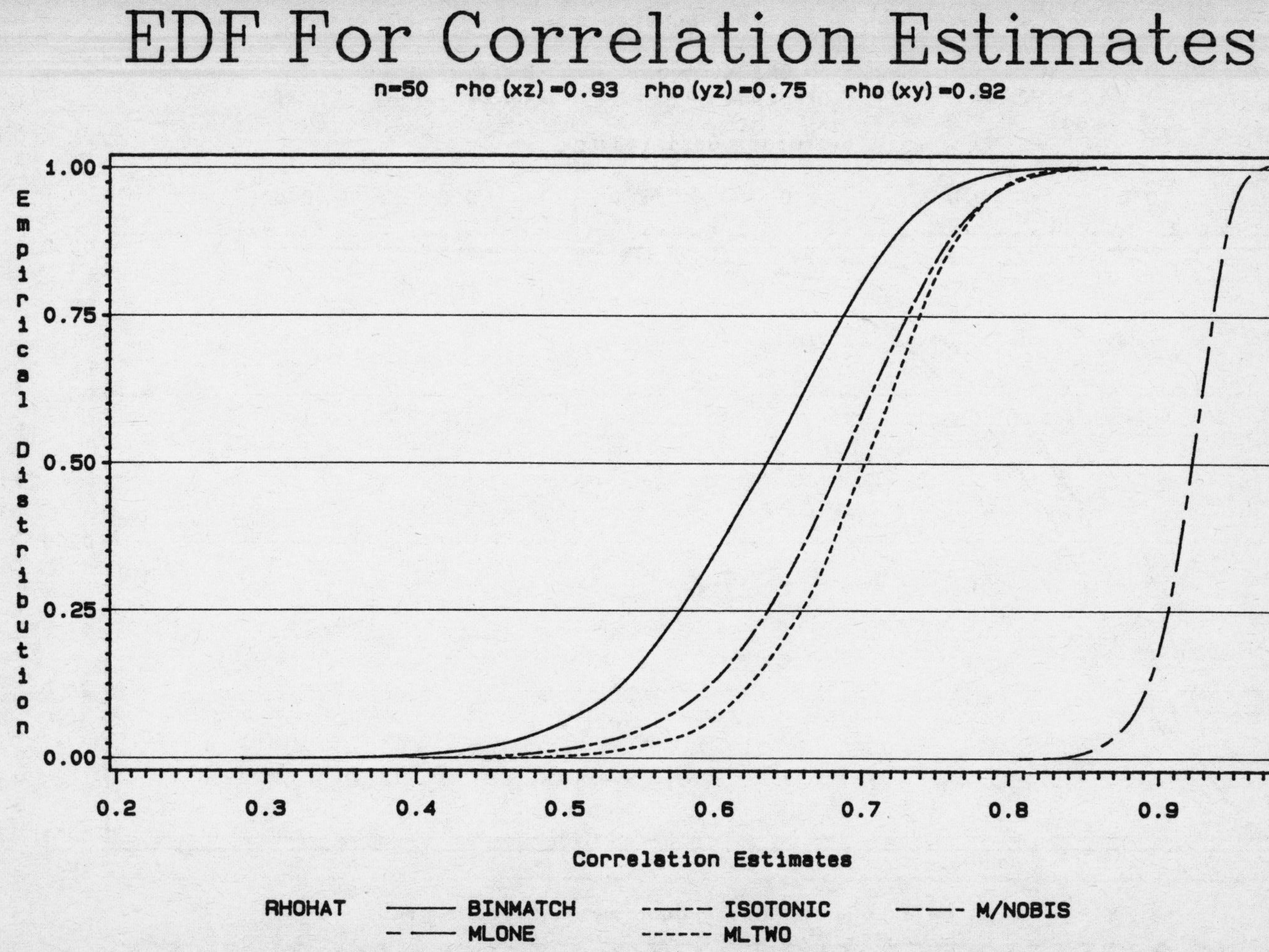
EDF For Correlation Estimates
n=50 rho (xz) =0.93 rho (yz) =0.75 rho (xy) =0.92
1.00
0.75
0.50
0.25
0.00
Empirical Distribution
0.2
0.3
0.4
0.5
0.6
0.7
0.8
0.9
1.0
Correlation Estimates
RHOHAT
BINMATCH
ISOTONIC
M/NOBIS
MLONE
MLTWO

Figure 3.27

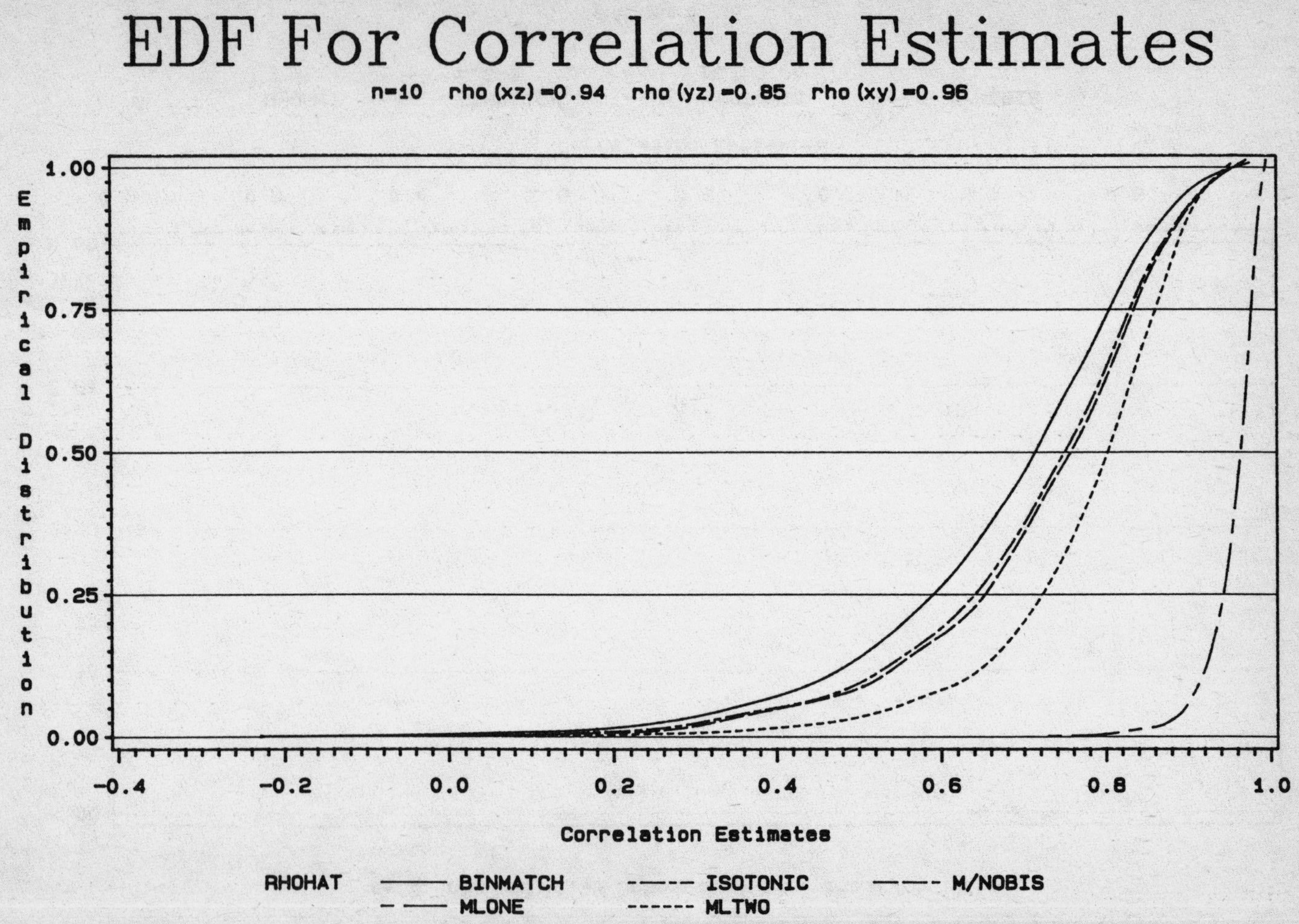
EDF For Correlation Estimates
n=10 rho (xz) =0.94 rho (yz) =0.85 rho (xy) =0.96
Empirical Distribution
1.00
0.75
0.50
0.25
0.00
-0.4
-0.2
0.0
0.2
0.4
0.6
0.8
1.0
Correlation Estimates
RHOHAT
BINMATCH
ISOTONIC
M/NOBIS
MLONE
MLTWO

Figure 3.28

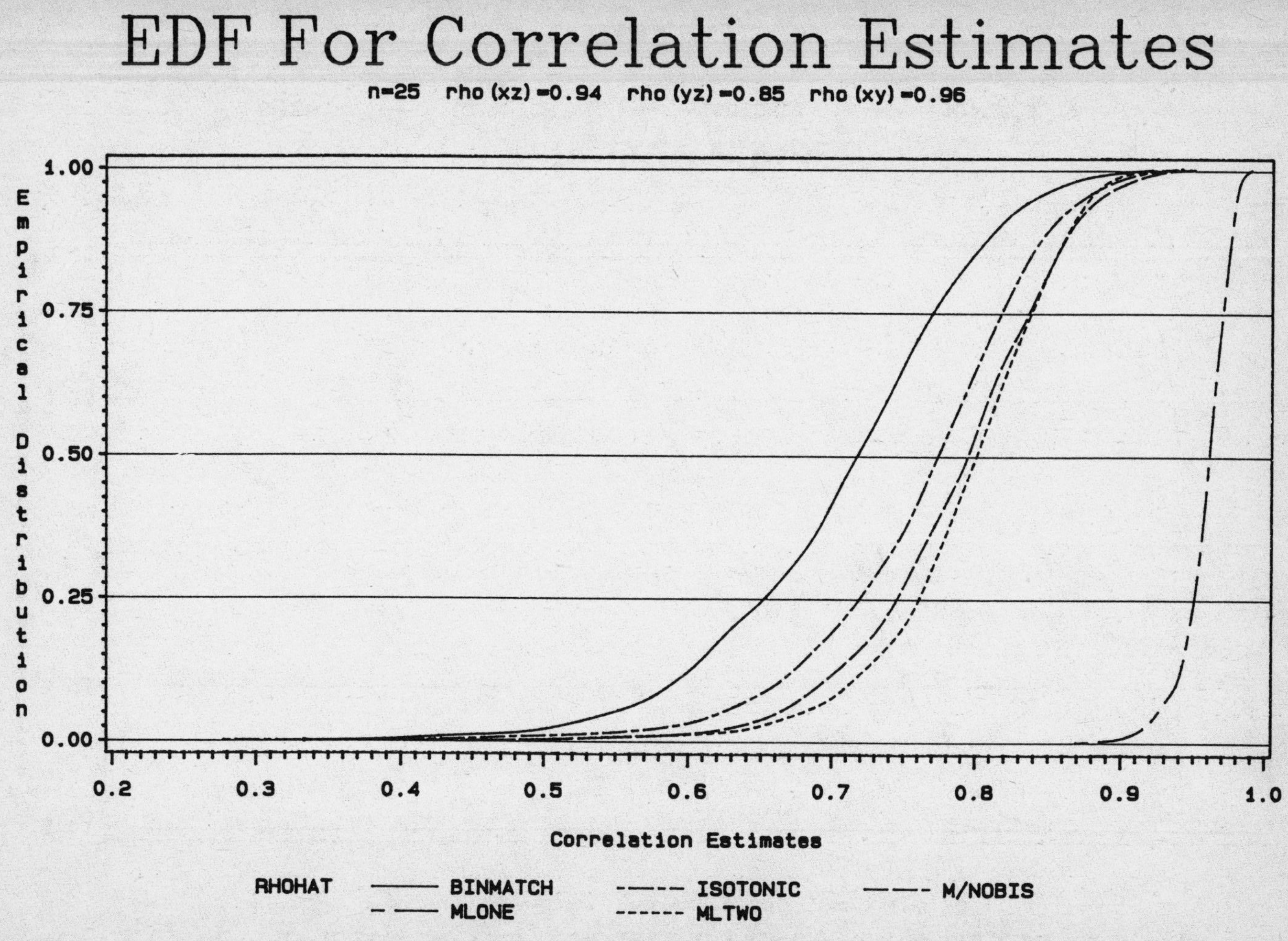
EDF For Correlation Estimates
n=25 rho (xz) =0.94 rho (yz) =0.85 rho (xy) =0.96
Empirical Distribution
1.00
0.75
0.50
0.25
0.00
0.2
0.3
0.4
0.5
0.6
0.7
0.8
0.9
1.0
Correlation Estimates
RHOHAT
BINMATCH
ISOTONIC
M/NOBIS
MLONE
MLTWO

Figure 3.29

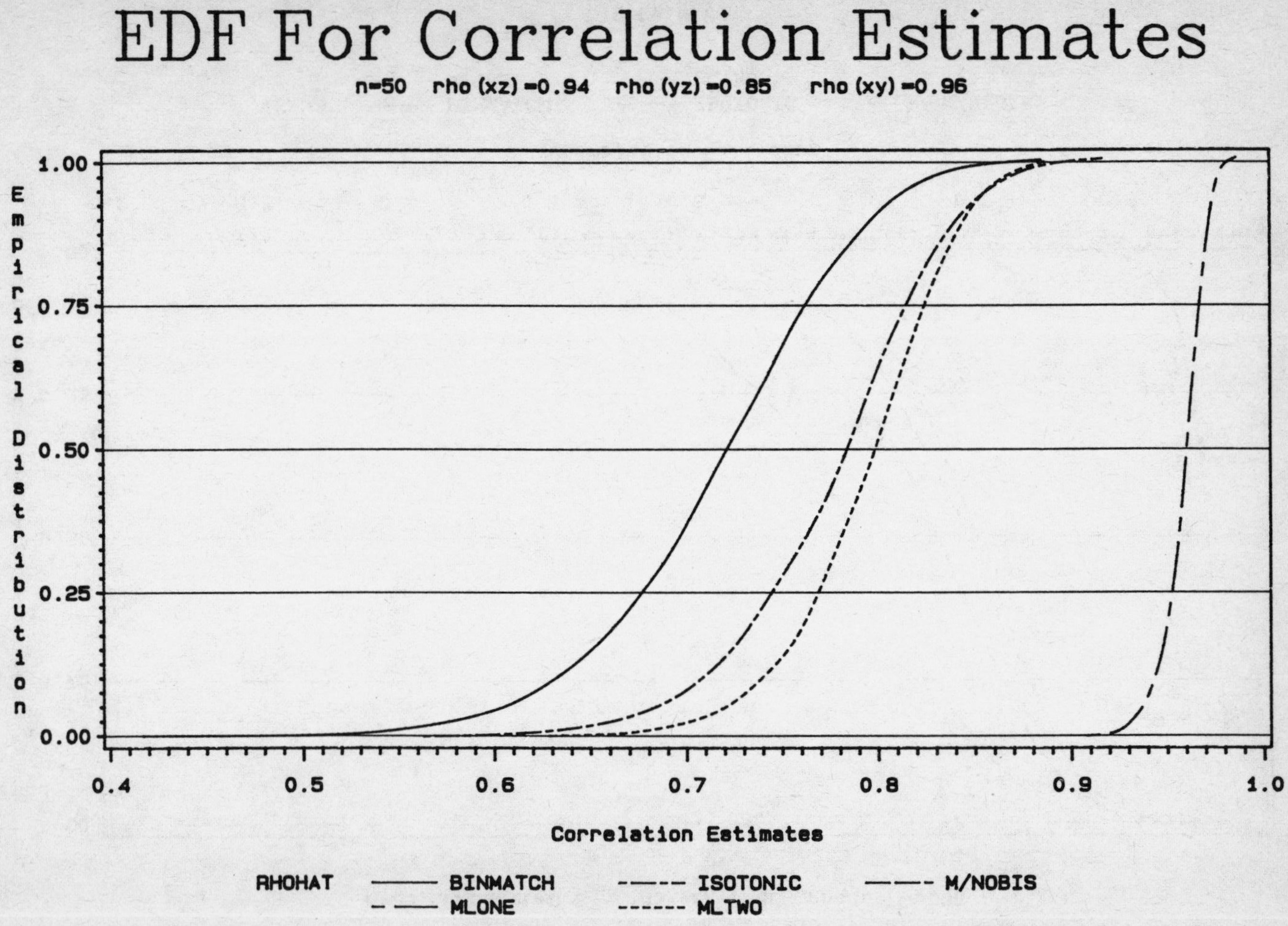
EDF For Correlation Estimates
n=50 rho (xz) =0.94 rho (yz) =0.85 rho (xy) =0.96
1.00
0.75
0.50
0.25
0.00
Empirical Distribution
0.4
0.5
0.6
0.7
0.8
0.9
1.0
Correlation Estimates
RHOHAT
BINMATCH
ISOTONIC
M/NOBIS
MLONE
MLTWO

Figure 3.30

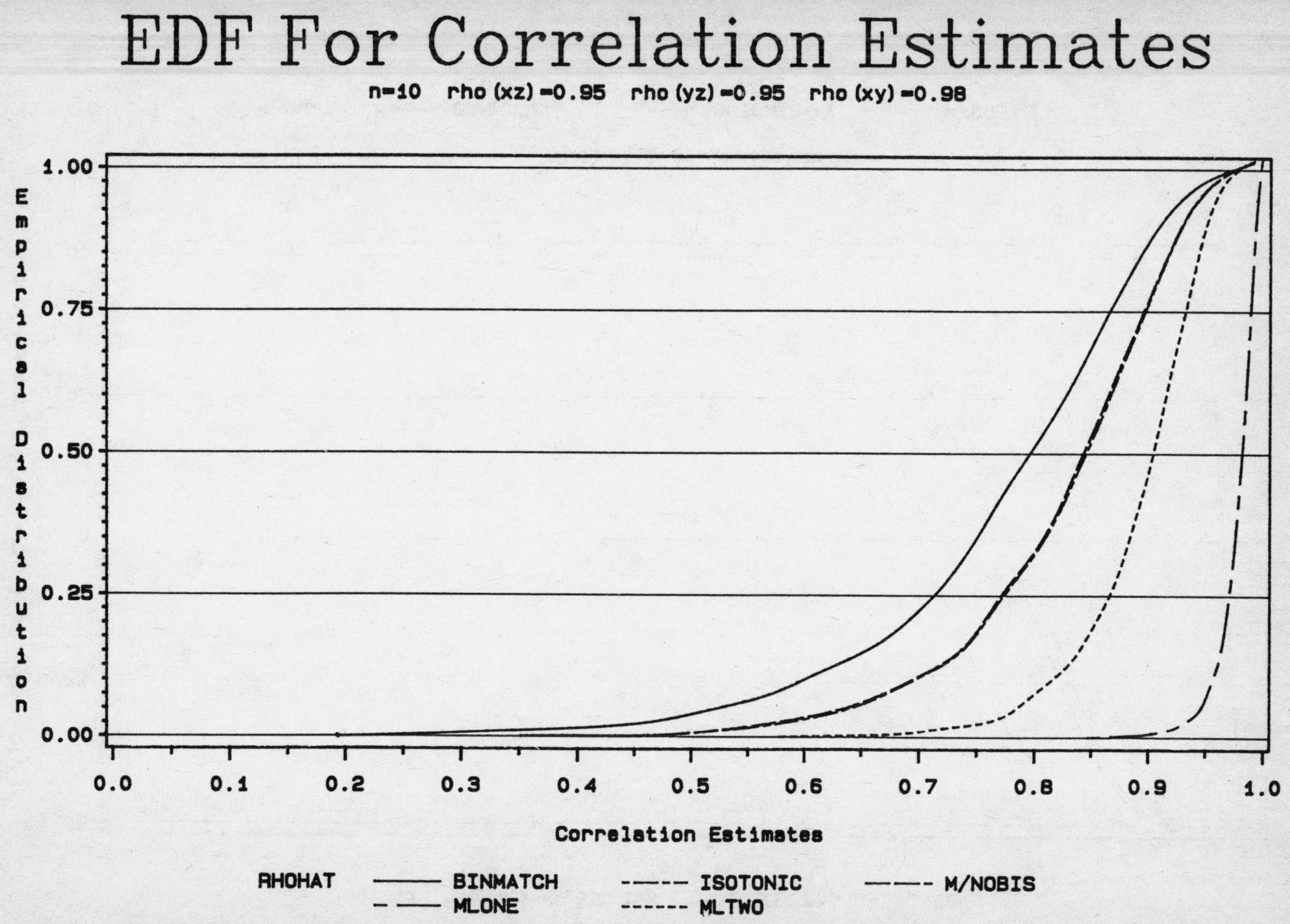

EDF For Correlation Estimates
n=10 rho (xz) =0.95 rho (yz) =0.95 rho (xy) =0.98
1.00
0.75
0.50
0.25
0.00
Empirical Distribution
0.0
0.1
0.2
0.3
0.4
0.5
0.6
0.7
0.8
0.9
1.0
Correlation Estimates
RHOHAT
BINMATCH
ISOTONIC
M/NOBIS
MLONE
MLTWO

Figure 3.31

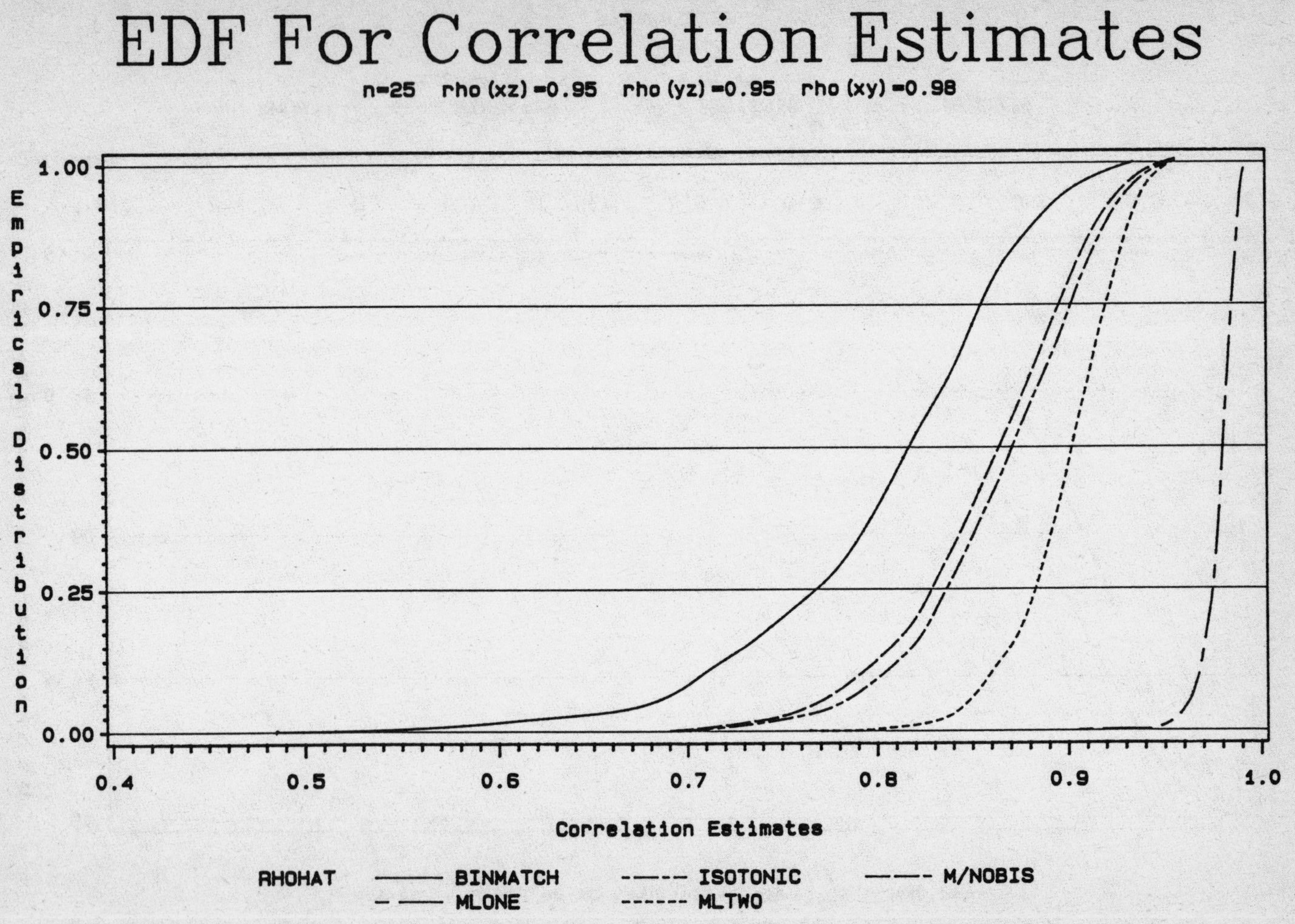
EDF For Correlation Estimates
n=25 rho (xz) =0.95 rho (yz) =0.95 rho (xy) =0.98
Empirical Distribution
1.00
0.75
0.50
0.25
0.00
0.4
0.5
0.6
0.7
0.8
0.9
1.0
Correlation Estimates
RHOHAT
BINMATCH
MLONE
ISOTONIC
MLTWO
M/NOBIS

Figure 3.32

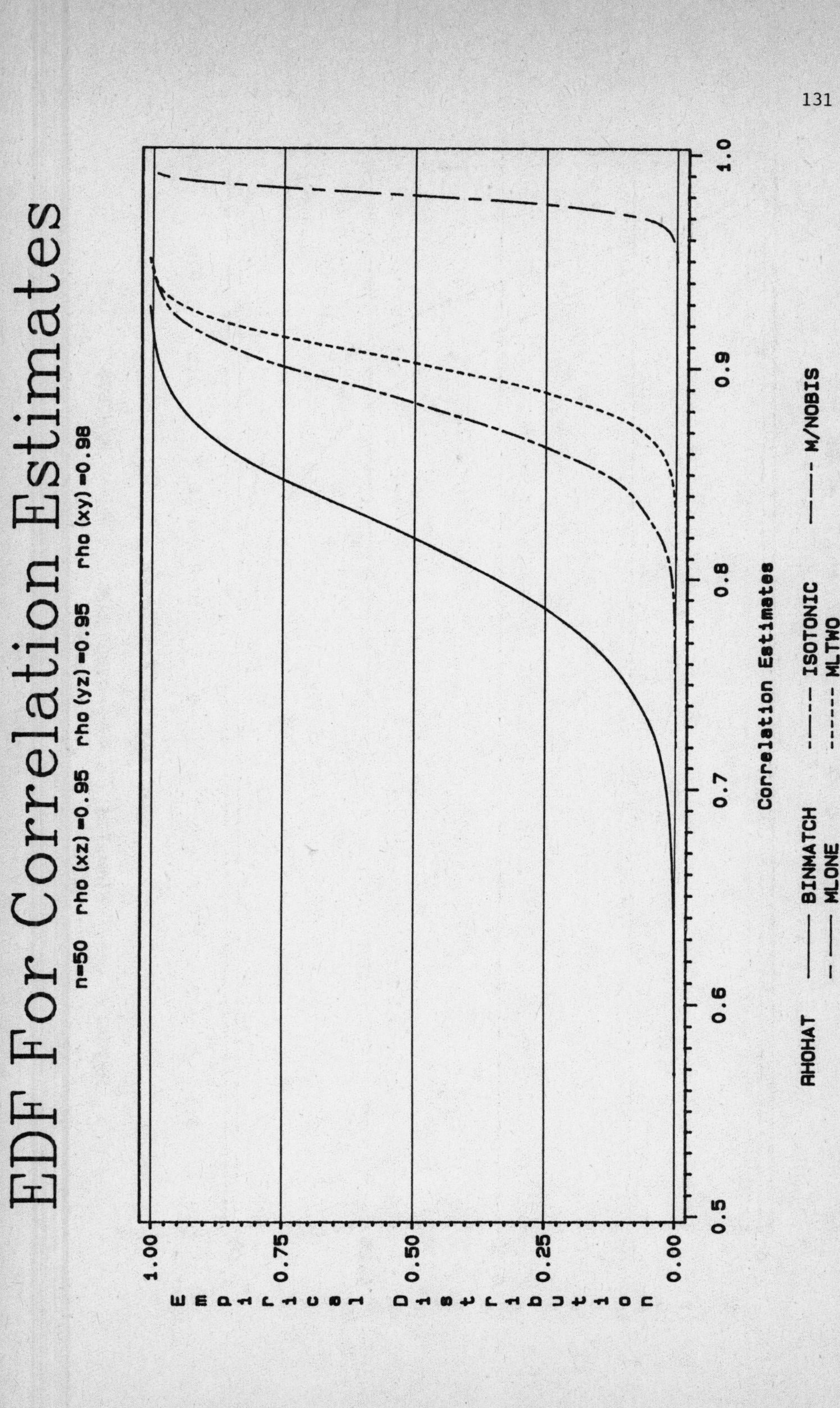
EDF For Correlation Estimates
n=50 rho (xz) =0.95 rho (yz) =0.95 rho (xy) =0.98
Empirical Distribution
1.00
0.75
0.50
0.25
0.00
0.5
0.6
0.7
0.8
0.9
1.0
Correlation Estimates
RHOHAT
BINMATCH
ISOTONIC
M/NOBIS
MLONE
MLTWO

Figure 3.33

EDF For Correlation Estimates

n=10 rho (xz) =0.97 rho (yz) =0.97 rho (xy) =0.99

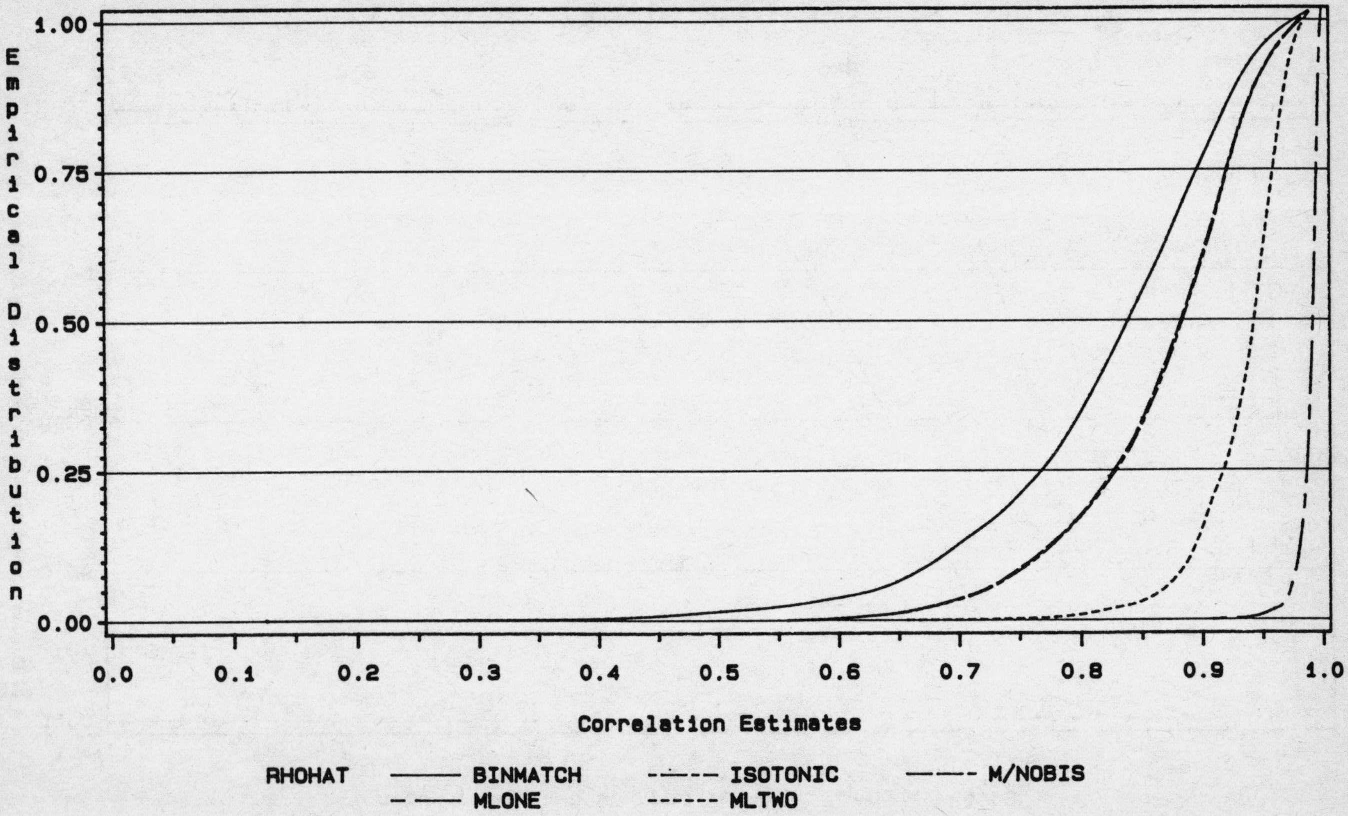

Figure 3.34

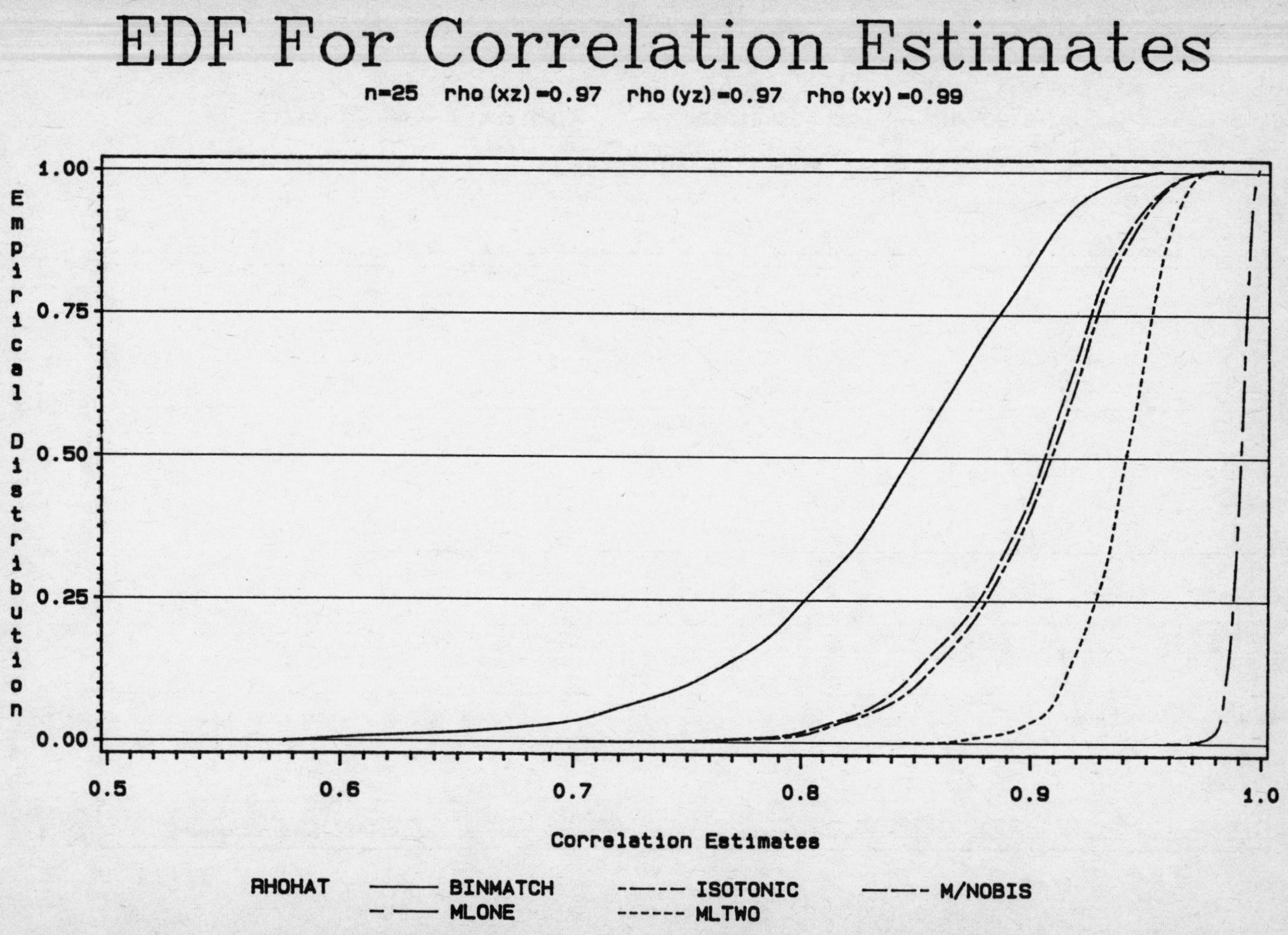

EDF For Correlation Estimates
n=25 rho (xz) =0.97 rho (yz) =0.97 rho (xy) =0.99
Empirical Distribution
1.00
0.75
0.50
0.25
0.00
0.5
0.6
0.7
0.8
0.9
1.0
Correlation Estimates
RHOHAT
BINMATCH
ISOTONIC
M/NOBIS
MLONE
MLTWO

Figure 3.35

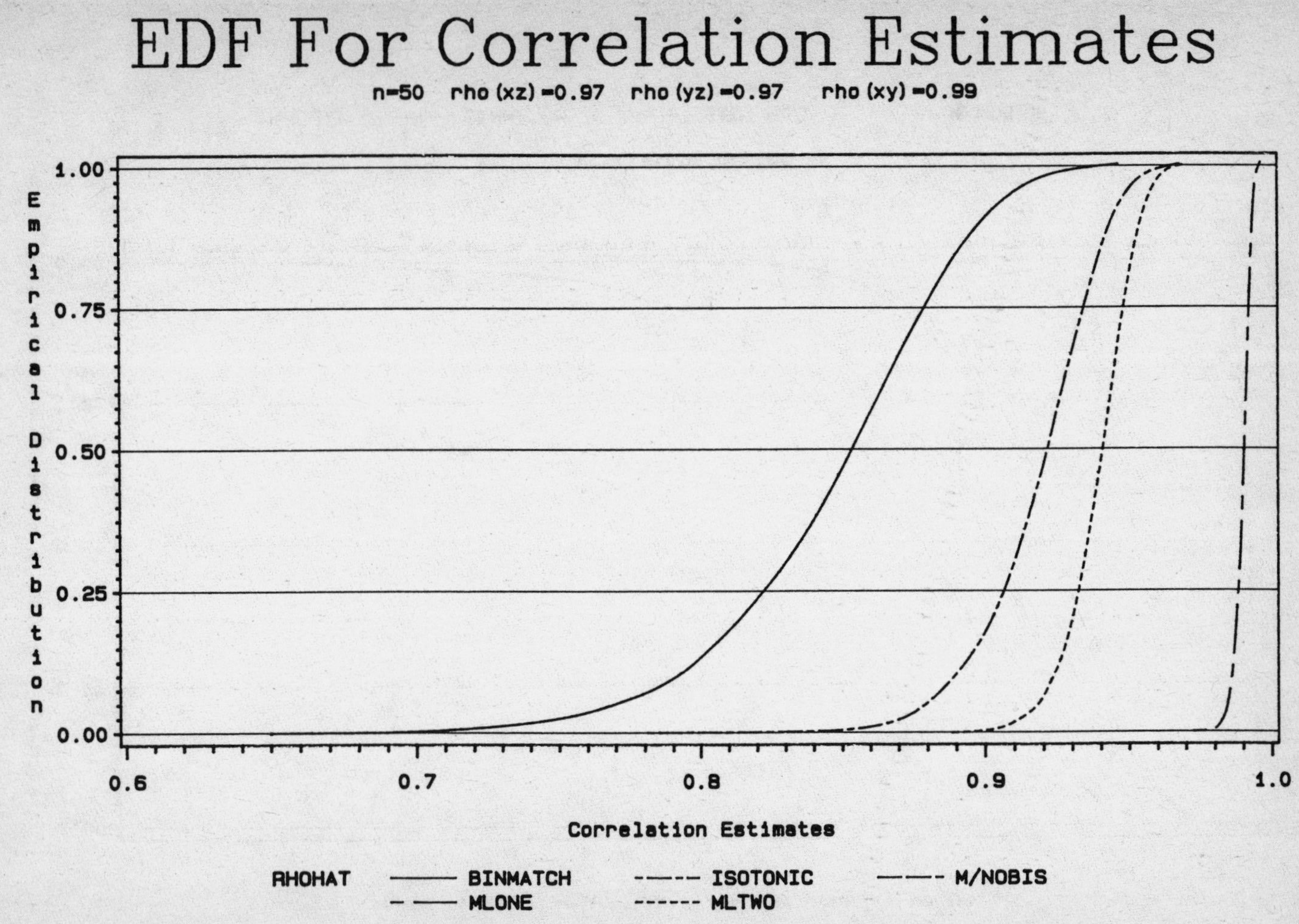

EDF For Correlation Estimates
n=50 rho (xz) =0.97 rho (yz) =0.97 rho (xy) =0.99
Empirical Distribution
1.00
0.75
0.50
0.25
0.00
0.6
0.7
0.8
0.9
1.0
Correlation Estimates
RHOHAT
BINMATCH
ISOTONIC
M/NOBIS
MLONE
MLTWO

Figure 3.36

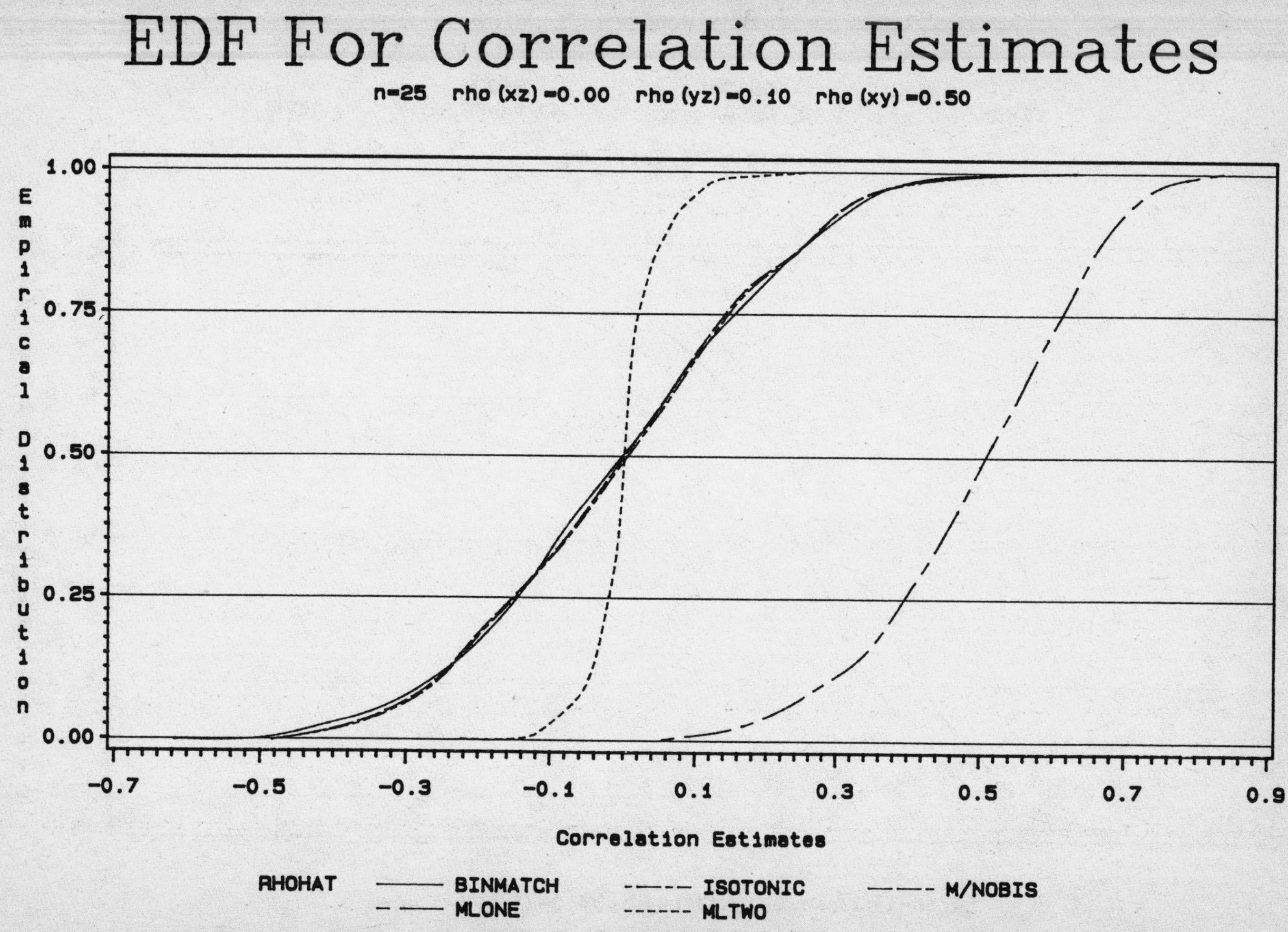
EDF For Correlation Estimates
n=25 rho (xz) =0.00 rho (yz) =0.10 rho (xy) =0.50
1.00
0.75
0.50
0.25
0.00
Empirical Distribution
-0.7
-0.5
-0.3
-0.1
0.1
0.3
0.5
0.7
0.9
Correlation Estimates
RHOHAT
BINMATCH
MLONE
ISOTONIC
MLTWO
M/NOBIS

Figure 3.37

EDF For Correlation Estimates

n=25 rho (xz) =0.92 rho (yz) =0.65 rho (xy) =0.75

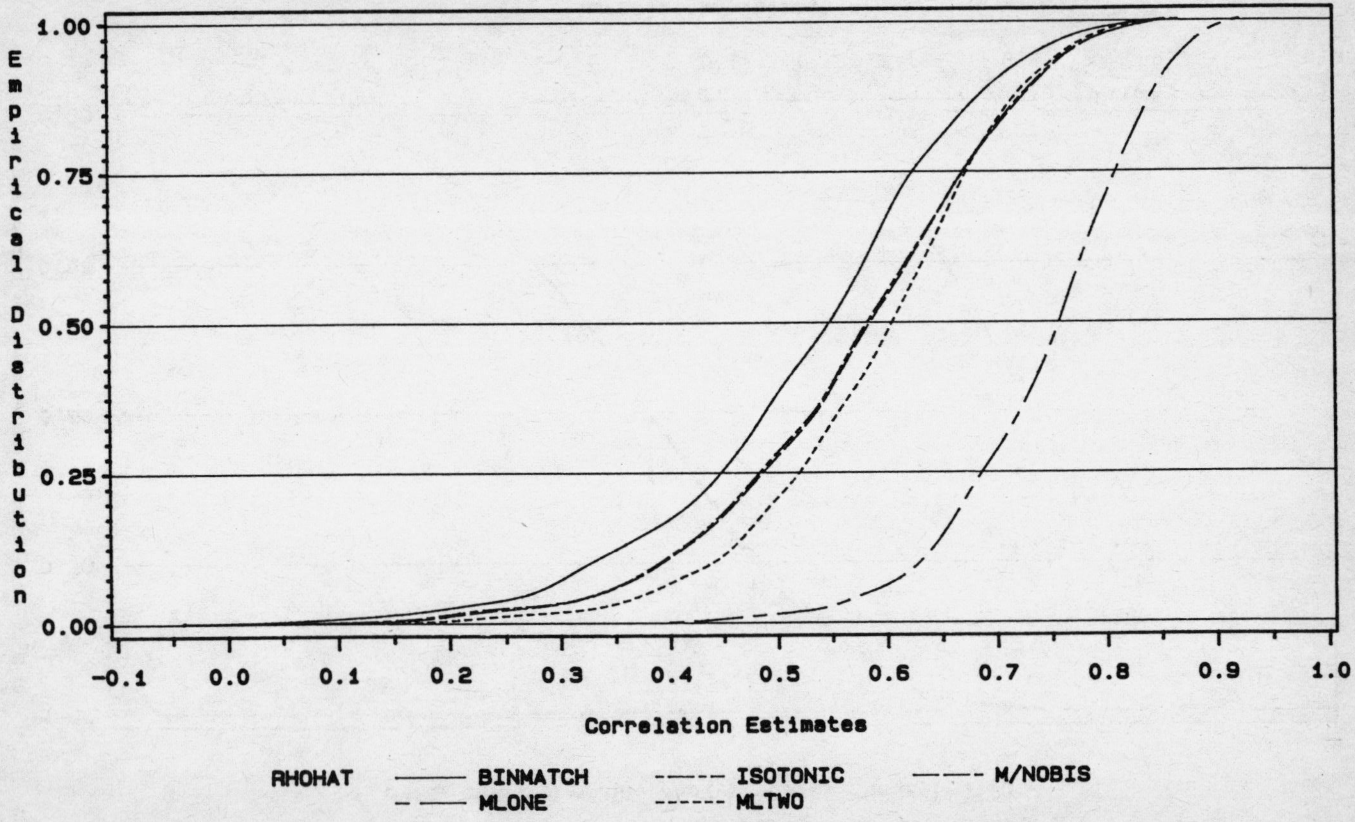

Figure 3.38

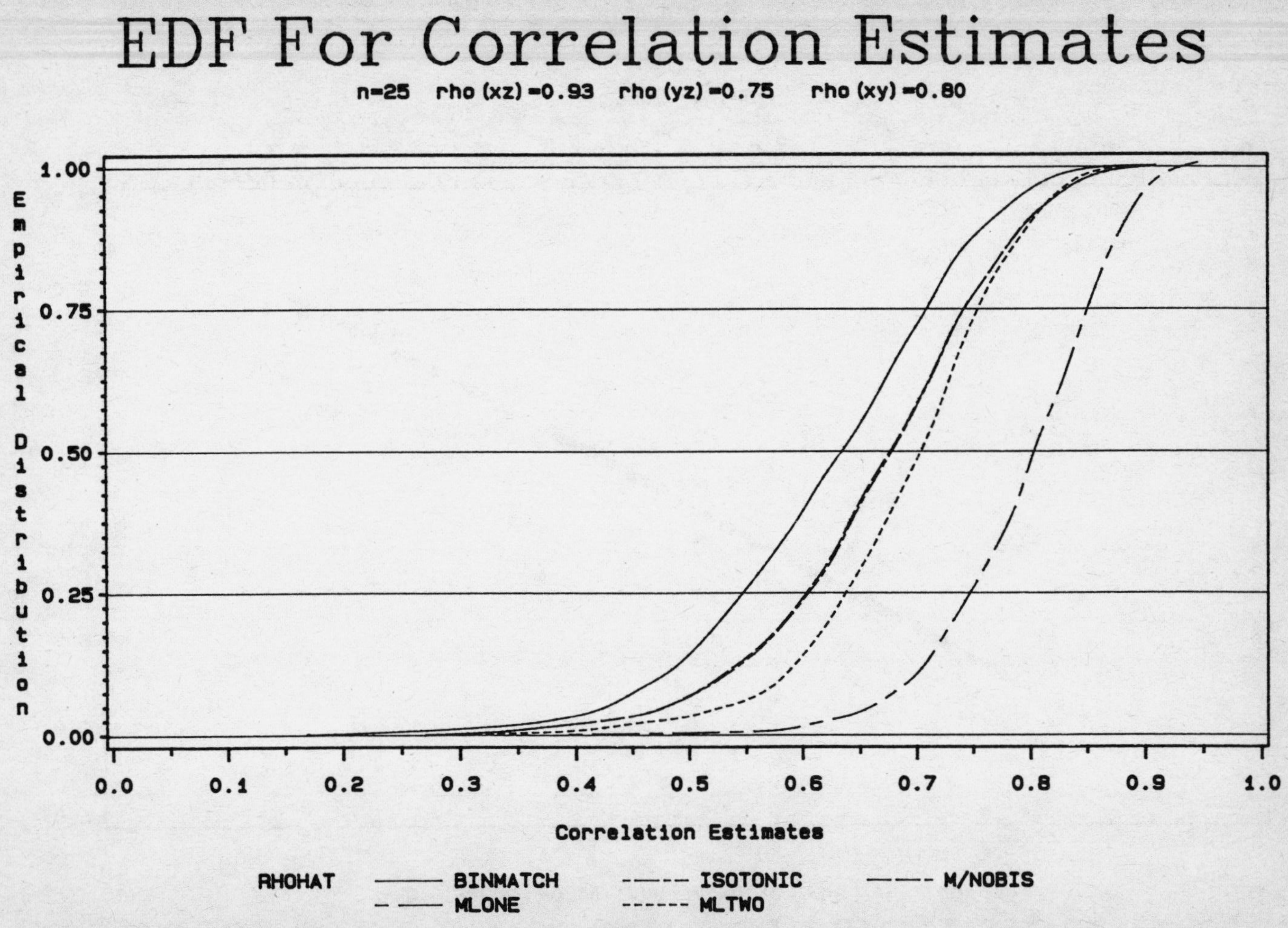
EDF For Correlation Estimates
n=25 rho (xz) =0.93 rho (yz) =0.75 rho (xy) =0.80
Empirical Distribution
1.00
0.75
0.50
0.25
0.00
0.0
0.1
0.2
0.3
0.4
0.5
0.6
0.7
0.8
0.9
1.0
Correlation Estimates
RHOHAT
BINMATCH
ISOTONIC
M/NOBIS
MLONE
MLTWO

Figure 3.39

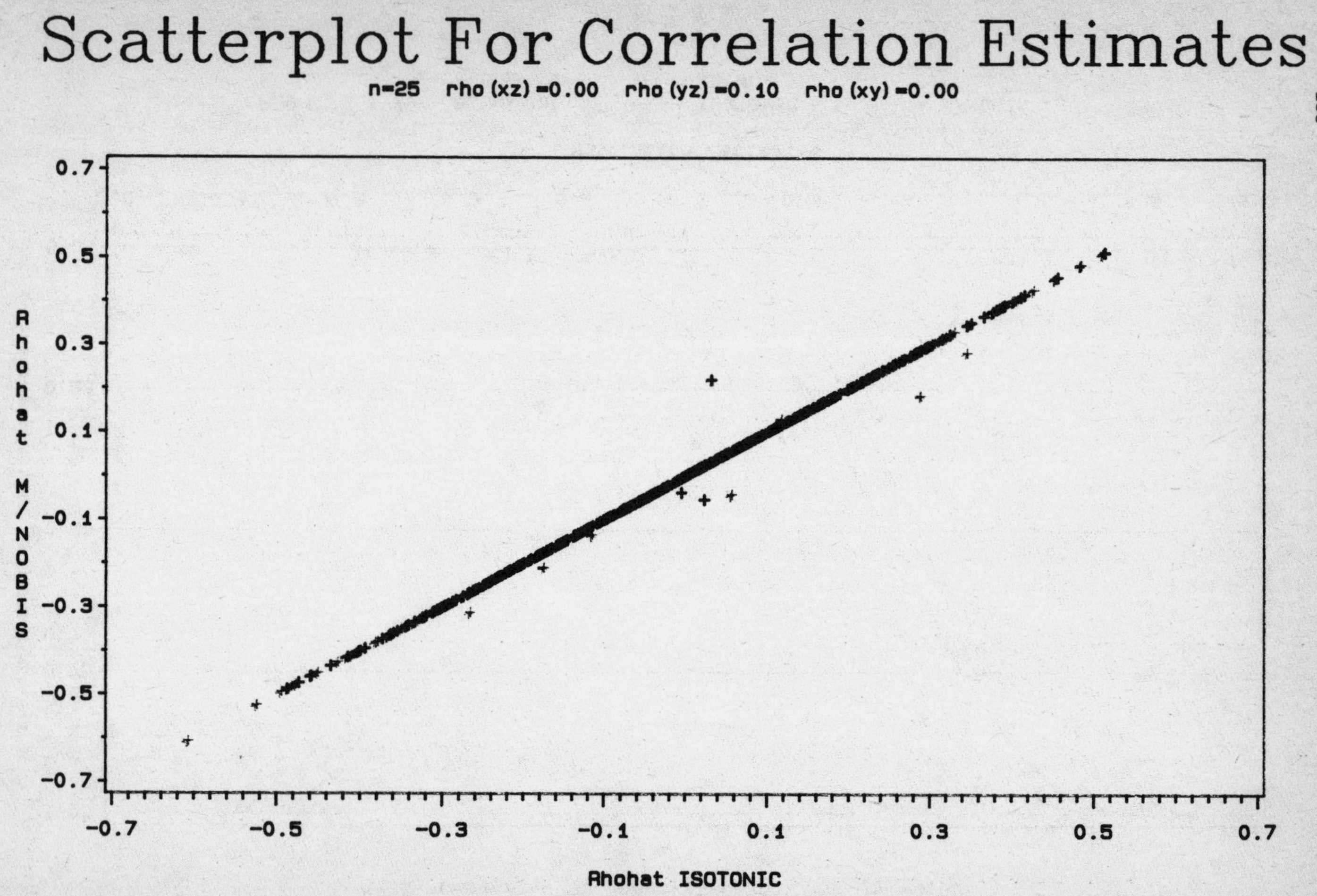
Scatterplot For Correlation Estimates
n=25 rho (xz) =0.00 rho (yz) =0.10 rho (xy) =0.00
Rhohat M/NOBIS
0.7
0.5
0.3
0.1
-0.1
-0.3
-0.5
-0.7
-0.7
-0.5
-0.3
-0.1
0.1
0.3
0.5
0.7
Rhohat ISOTONIC

Figure 3.40

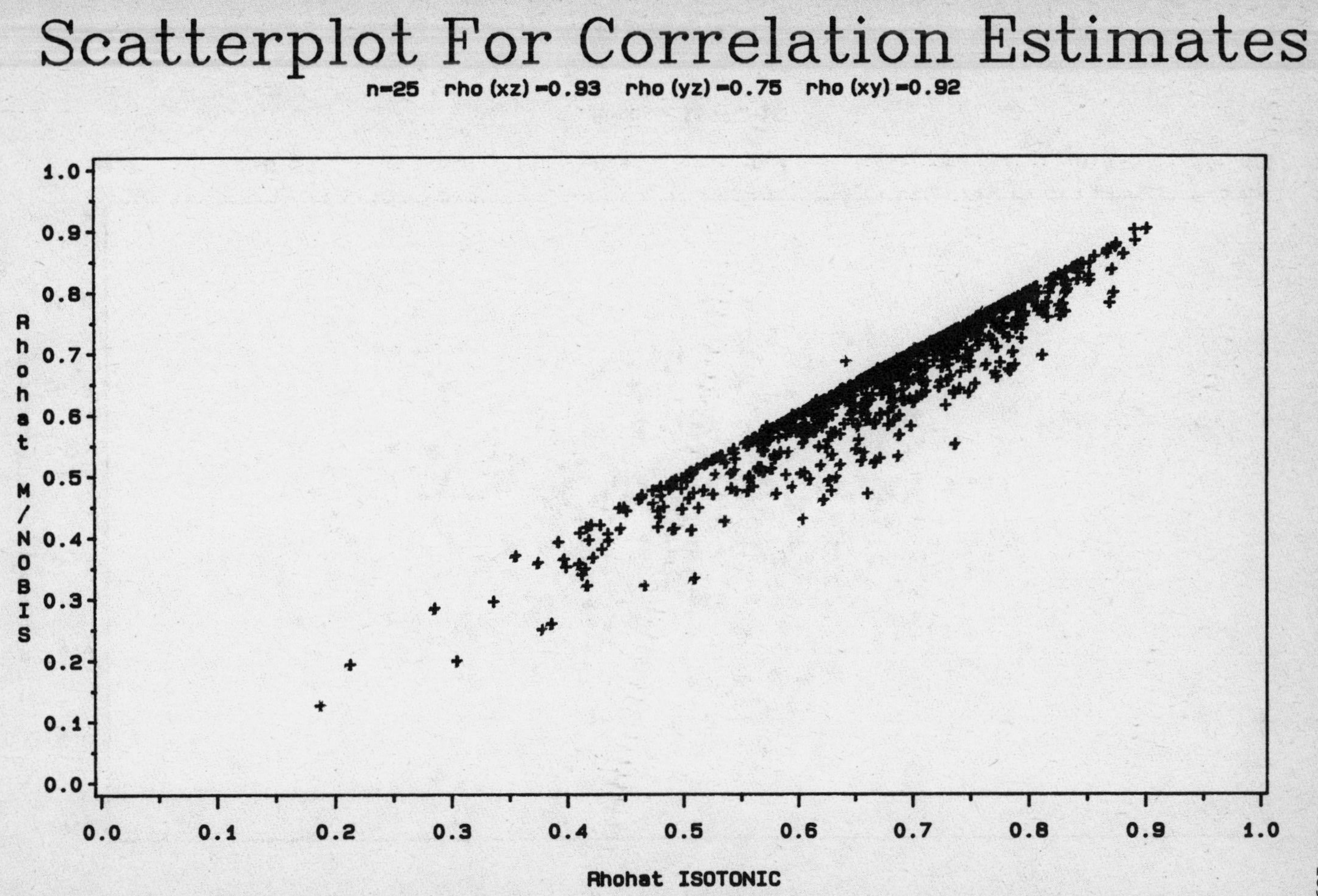
Scatterplot For Correlation Estimates
n=25 rho (xz)=0.93 rho (yz)=0.75 rho (xy)=0.92
1.0
0.9
0.8
0.7
0.6
0.5
0.4
0.3
0.2
0.1
0.0
Rhohat M/NOBIS
0.0
0.1
0.2
0.3
0.4
0.5
0.6
0.7
0.8
0.9
1.0
Rhohat ISOTONIC

Figure 3.41

Scatterplot For Correlation Estimates

n=25 rho (xz) =0.00 rho (yz) =0.10 rho (xy) =0.00

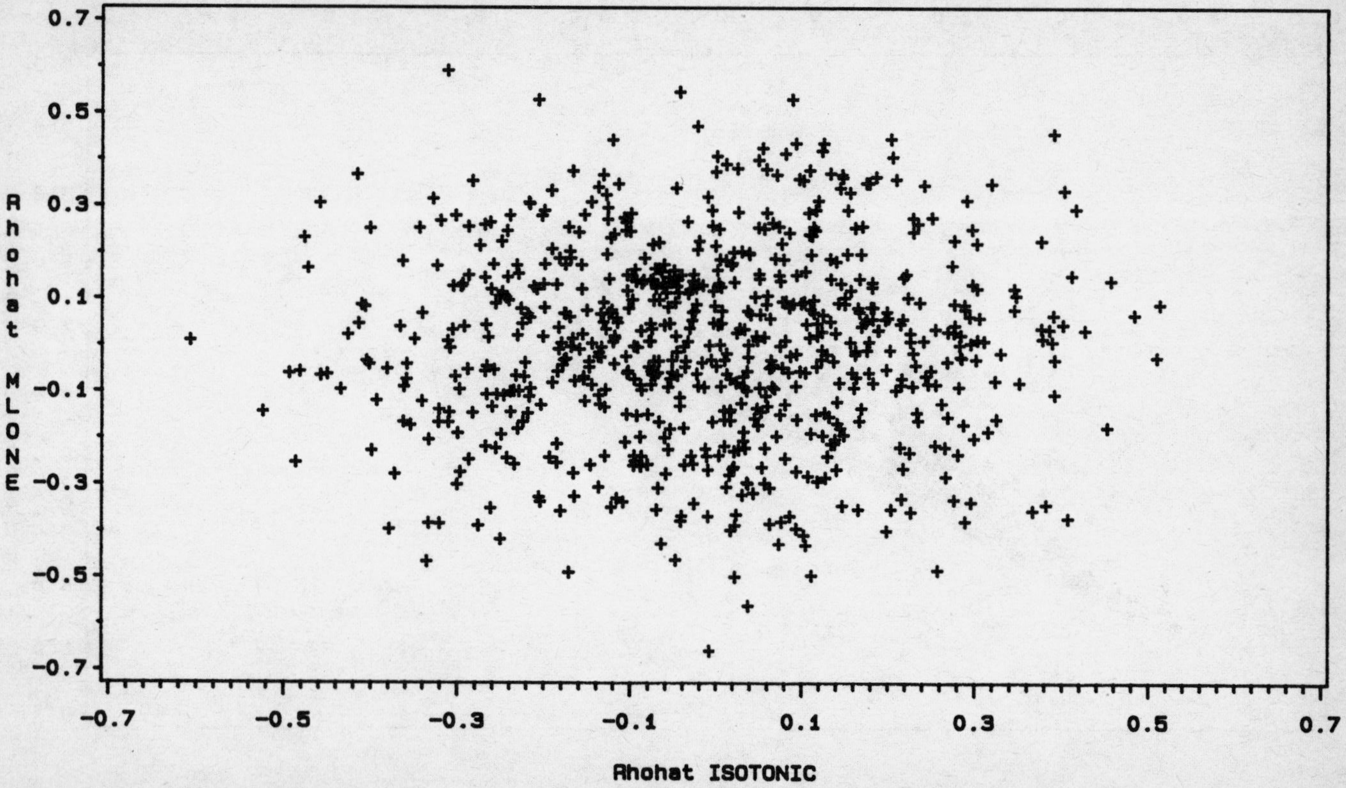

Figure 3.42

Scatterplot For Correlation Estimates

n=25 rho (xz) =0.00 rho (yz) =0.10 rho (xy) =0.00

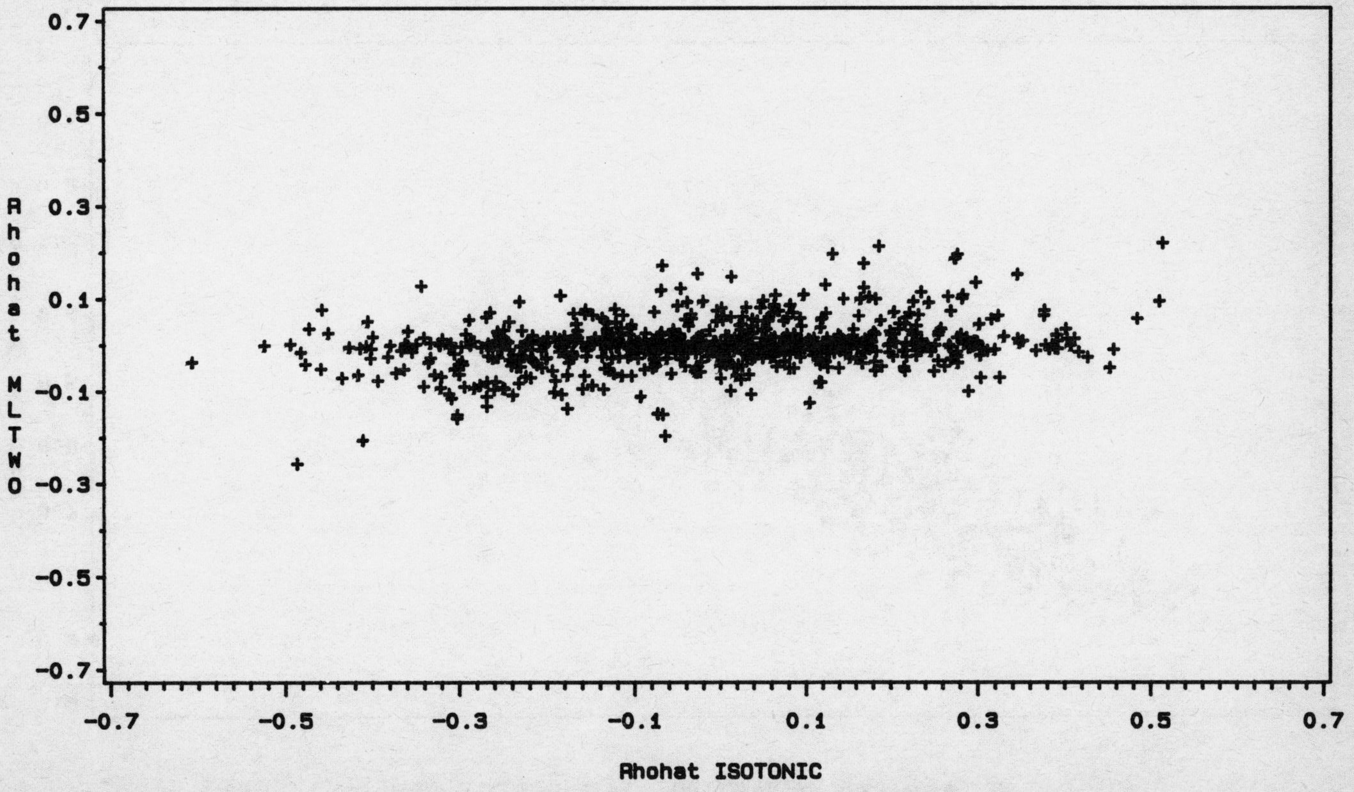

Figure 3.43

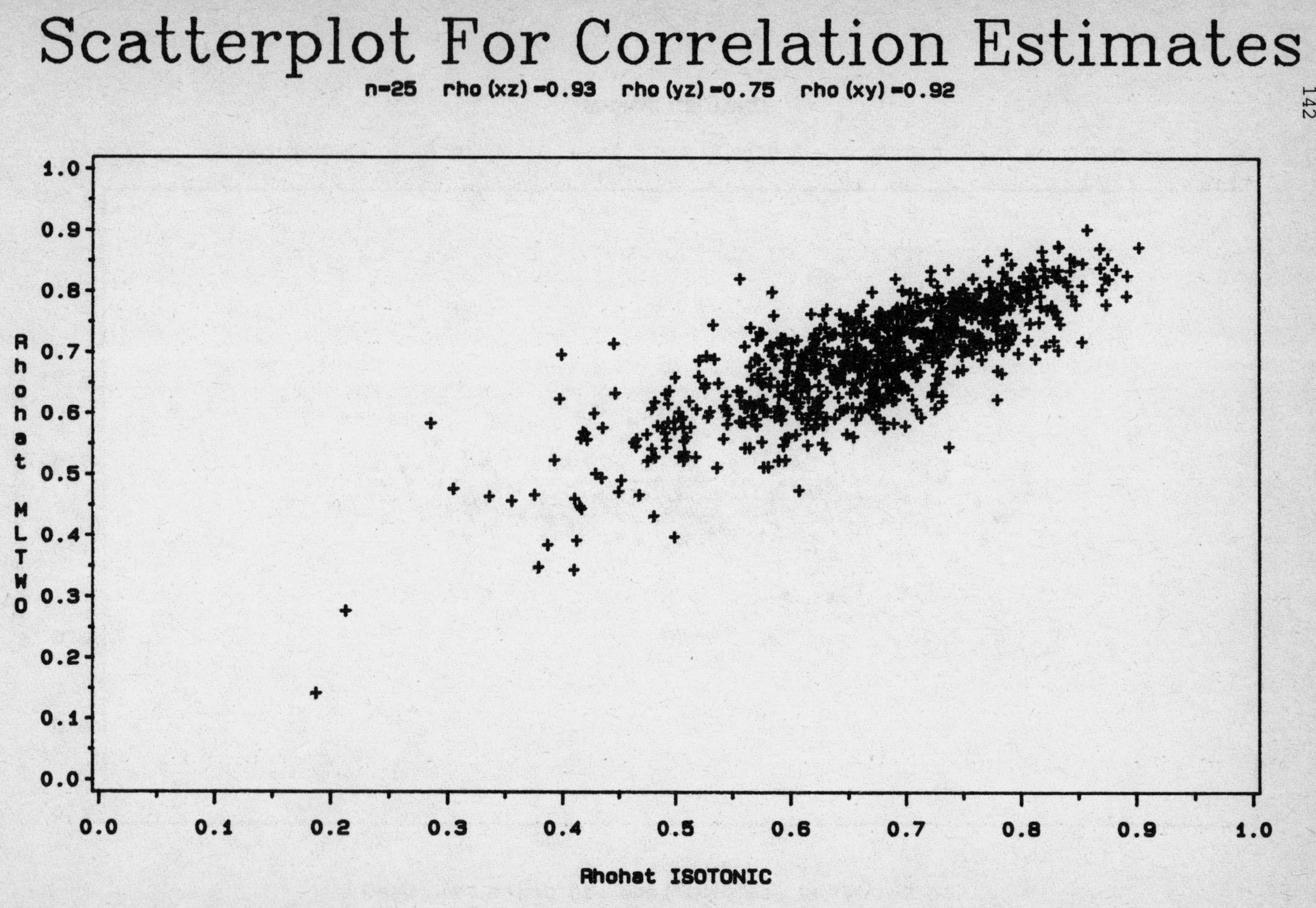
Scatterplot For Correlation Estimates
n=25 rho (xz) =0.93 rho (yz) =0.75 rho (xy) =0.92
Rhohat MLTWO
1.0
0.9
0.8
0.7
0.6
0.5
0.4
0.3
0.2
0.1
0.0
0.0 0.1 0.2 0.3 0.4 0.5 0.6 0.7 0.8 0.9 1.0
Rhohat ISOTONIC

Figure 3.44

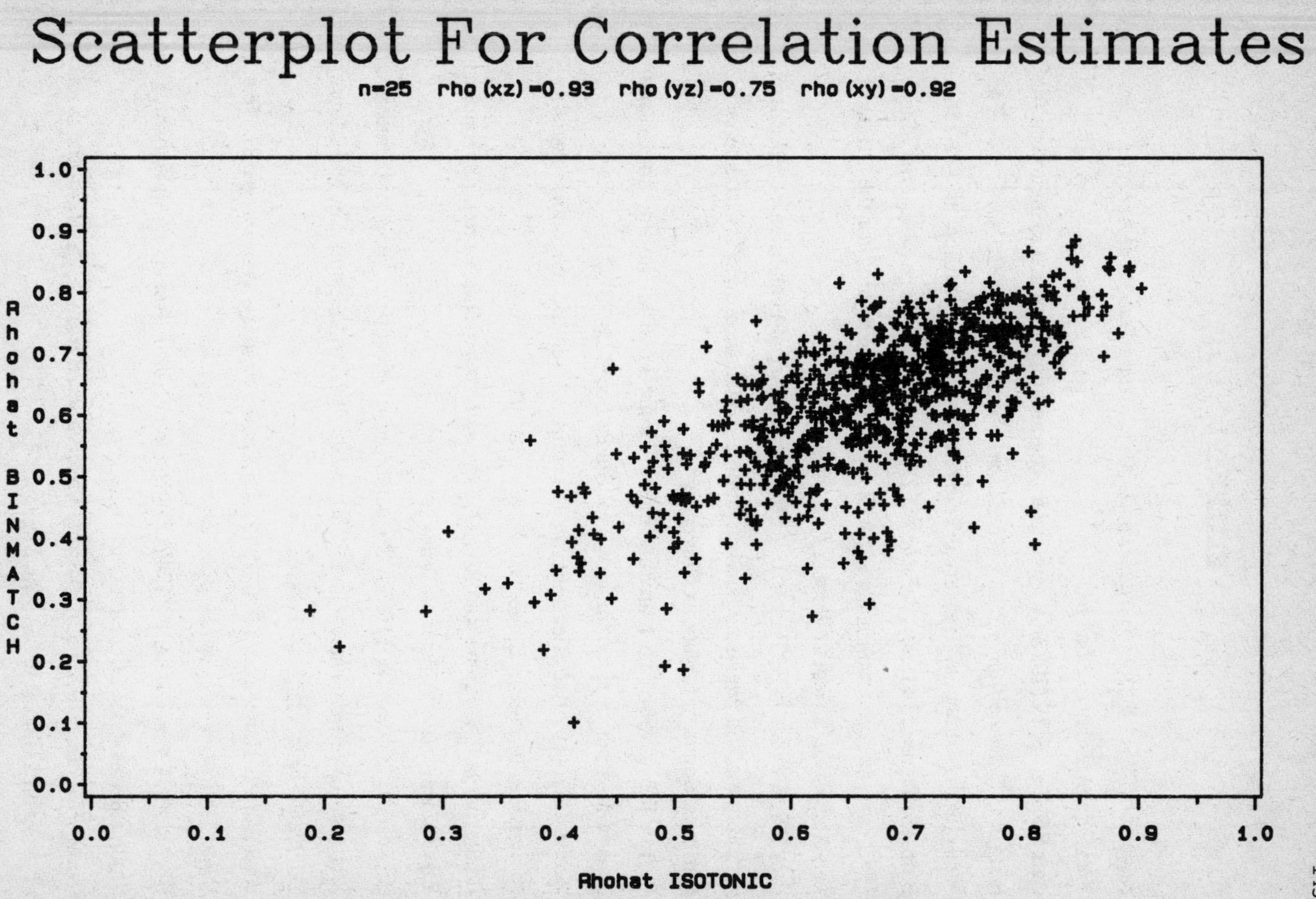
Scatterplot For Correlation Estimates
n=25 rho (xz) =0.93 rho (yz) =0.75 rho (xy) =0.92
Rhohat BINMATCH
1.0
0.9
0.8
0.7
0.6
0.5
0.4
0.3
0.2
0.1
0.0
0.0 0.1 0.2 0.3 0.4 0.5 0.6 0.7 0.8 0.9 1.0
Rhohat ISOTONIC

Figure 3.45

REFERENCES

Anderson TW (1984) *An Introduction to Multivariate Statistical Analysis.* New York: John Wiley & Sons Inc

Barr RS, Turner JS (1978) A new linear programming approach to microdata file merging, *Compendium of Tax Research.* Washington DC: Office of Tax Analysis Department of Treasury US Government Printing Office: p. 131-158

Barr RS, Turner JS (1980) *Merging the 1977 Statistics of Income and the March 1978 current population survey.* Washington, DC: Prepared for the Office of Tax Analysis, US Dept of Treasury

Barr RS, Stewart WH, Turner JS (1982) An empirical evaluation of statistical matching methodologies, unpublished mimeo. Dallas: Edwin L Cox School of Business, Southern Methodist University

Barton DE (1958) The matching distributions: Poisson limiting forms and derived methods of approximation. J R Statist Soc B 20:73-92

Beebe GW (1979) Reflections on the work of the atomic bomb casuality commission in Japan. Epidemiol. Rev. 1: 184-210

Bhattacharya RN, Ranga Rao R (1976) *Normal Approximation and Asymptotic Expansions.* New York: John Wiley and Sons

Bickel PJ, Yahav JA (1977) On selecting a subset of good populations. *Statistical Decision Theory and Related Topics II,* (ed.) Gupta SS, and Moore DS, New York: Academic Press

Bleistein PJ, Handelsman RA (1975) *Asymptotic expansions of integrals.* New York: Holt, Rinehart, Winston

Bochner S, Chandrasekar K (1949) *Fourier Transforms* Annals of Mathematical Studies Series #19. Princeton University Press

Chew MC Jr, (1973) On pairing observations from a distribution with monotone likelihood ratio. Ann Statist 1:433-445

Chow YS and Teicher H (1978) *Probability Theory: Independence Interchangeability Martingales,* New York: Springer-Verlag

Chung KL (1974) *A Course in Probability Theory, 2nd edn.* New York: Academic Press

Dawber TR et al (1963) An approach to the longitudinal studies in a community: The Framingham study. Ann NY Acad Sci 107:539-556.

David FN, Barton DE (1962) *Combinatorial Chance.* London: Griffin

Dawid AP (1979) Conditional independence in statistical theory. JR Statist Soc B 41:1-31

DeGroot MH (1980) Optimal Matching: A Survey. *Symposia Mathematica,* vol XXV. Instituto Nazionale di Alta Matematica, Francesco Saveri, Roma. New York: Academic Press: p. 123-135

DeGroot MH (1987) Record linkage and matching systems. In: *Encyclopedia of Statistical Sciences* (ed) Kotz S, Johnson NL: vol 7:649-654

DeGroot MH, Feder PI, Goel PK (1971) Matchmaking. Ann Math Statist 42:578-593

DeGroot MH, Goel PK (1976) The matching problem for multivariate normal data. Sankhya Ser B 38:14-29

DeGroot MH, Goel PK (1980) Estimation of the correlation coefficient from a broken random sample. Ann Statist 8:264-278

Dubois NS, D'Andrea (1969) A solution to the problem of linking multivariate documents. J Amer Statist Assn 64:163-174

Eaton M, Kariya T (1983) Multivariate tests with incomplete data. Ann Statist 11:654-665

Esary JD, Proschan F, Walkup DW (1967) Association of random variables with applications. Ann Math Statist 38:1466-1474

Fellegi IP, Sunter AB (1969) A theory of record linkage. J Amer Statist Assc 64:1183-1210

Fellegi IP (1978) *Discussion* 1977 proceedings of the American Statistical Association, social statistics section: p. 762-764

Feller W (1968) *An Introduction to Probability Theory and Its Applications, 3rd eds: vol I.* New York: John Wiley and Sons

Fishburn PC, Doyle PG, Shepp LA (1988) The match set of a random permutation has the FKG property. Ann Prob 16:1194-1214

Goel PK (1975) On re-pairing observations in a broken random sample. Ann Stat 3:1364-1369

Goel PK, Ramalingam T (1985) The matching methodology: some statistical properties, Tech Report #333, Department of Statistics, The Ohio State University

Goel PK, Ramalingam T (1987) Some properties of the maximum likelihood strategy for re-pairing a broken random sample. J Statist Plann Inference 16:237-248

Hollenbeck K, Doyle P (1979) Distributional characteristics of a merged microdata file. American Statistical Association proceedings of the section on survey research methods, p. 418-420

Kadane JB (1978) Some statistical problems in merging data files. *Compendium of Tax Research*, Washington DC: Office of Tax Analysis, Department of Treasury. US Government Printing Office: p. 159-179

Kelley RP (1983) A preliminary study of the error structure of statistical matching. American Statistical Association proceedings of the section on social statistics, p 206-208

Kilss B, Alvey W (editors) (1985) Record linkage techniques. Proceedings of the workshop on exact matching methodologies, Arlington, Virginia

Lancaster HO (1969) *The Chi-Squared Distribution.* New York: John Wiley and Sons

Lehmann EL (1966) Some concepts of dependence. Ann Math Statist 37:1137-1153

Mardia KV (1970) *Families of Bivariate Distributions* #27. Griffin's Statistical Monographs and Courses (ed) Alan Stuart. London: Charles Griffin & Co

Montmort PR de (1708) Essay d'analyse sur les jeux des hazards, 1st ed. Paris

Newman CM (1982) Asymptotic independence and limit theorems for positively and negatively dependent random variables. *Inequalities in Statistics and*

Probability (ed Tong YL), Lecture Notes monograph series, vol V, Hayward, CA: Institute of Mathematical Statistics

Okner B (1972) Constructing a new data base from existing micro data sets: 1966 Merge File. Annals of Economic and Social Measurement, 1, p 325-342

Radner et al (1980) Report on exact and statistical matching techniques, *Statistical Policy Working Paper 5*. Office of Federal Statistical Policy and Standards, US Dept of Commerce

Ramalingam T (1985) Statistical Properties of the file-merging methodology. Ph.D. Thesis, Purdue University

Randles RH, Wolfe DA (1979) *Introduction to the Theory of Nonparametric Statistics*. New York: John Wiley and Sons

Rodgers WL (1984) An evaluation of statistical matching. Journal of Business and Economic Statistics 2:91-102

Rubin DB (1986) Statistical matching using file concatenation with adjusted weights and multiple imputations. Journal of Business and Economic Statistics 4:87-94

Schweizer B, Wolff EF (1981) On nonparametric measures of dependence for random variables. Ann Statist 9:879-885

Serfling RJ (1980) *Approximation Theorems of Mathematical Statistics* . New York: John Wiley & Sons Inc

Shaked M (1979) Some concepts of positive dependence for bivariate interchangeable distributions. Ann Inst Statist Math 31 Part A:67-84

Sims CA (1972) Comments (On Okner 1972). Annals of Economic and Social Measurement 1:343-345

Sims CA (1978) Comments (On Kadane 1978) *1978 Compendium of Tax Research*, Washington DC: Office of Tax Analysis, Dept of the Treasury. US Govt Printing Office: p. 172-177

Srivastava MS, Khatri CG (1979) *An Introduction to Multivariate Statistics*. New York: Elsevier North Holland

Stein C (1986) Approximate computation of expectations, *Lecture Notes - Monograph Series*, Vol VII, Hayward, CA: Institute of Mathematical Statistics

Tong YL (1980) *Probability Inequalities in Multivariate Distributions.* New York: Academic Press

Widder DV (1941) *The Laplace Transform.* Princeton University Press

Woodbury MA (1983) Statistical record matching for files, *Incomplete Data in Samples Surveys, vol 3* Madow WH et al (eds) p. 173-202

Yahav JA (1982) On matchmaking, *Statistical Decision Theory and Related Topics III, vol 2* (eds) Gupta SS, Berger JO. New York: Academic Press, p. 497-504

Zionts S (1974) *Linear and Integer Programming.* Englewood Cliffs, NJ: Prentice-Hall

Zolutuchina LA, Latishev KP (1978) Asymptotic behavior of the expected number of coincidences of elements in a sequence of bivariate samples (in Russian). Leningrad Older Matimal Inst, Akad Nauk SSSR 79: 4-10

Author Index

Anderson, T. W.	81, 82, 84
Barr, R. S.	4, 9, 76, 86
Barton, D. E.	17
Beebe, G. W.	3
Bhattacharya, R. N.	60, 65
Bickel, P. J.	43
Bleistein, P. J.	48
Bochner, S.	65
Chandrasekar, K.	65
Chew, M. C.	19, 22, 39
Chow, Y. S.	44, 56
Chung, K. L.	44
David, F. N.	17
Dawber, T. R.	2
Dawid, A. P.	79
DeGroot, M. H.	2, 4, 8, 12, 14, 17, 18, 19, 20, 21, 22, 24, 77
Doyle, P.	9
Dubois, N. S.	2
Eaton, M. L.	81
Esary, J. D.	51
Feder, P. I.	12, 14, 17, 19, 20, 21
Fellegi, I. P.	2
Feller, W.	16, 17
Fishburn, P. C.	53, 56
Goel, P. K.	12, 14, 17, 18, 19, 20, 21, 22, 24, 40, 49, 53, 77
Handlesman, R. A.	48
Hollenbeck, K.	9
Kadane, J. B.	4, 13, 68, 69, 71, 76, 86
Kariya, T.	81
Kelley, R. P.	12
Khatri, C. G.	82
Kilss, B.	3
Lancaster, H. O.	33, 64
Latishev, K. P.	48, 49, 64
Lehmann, E. L.	50
Mardia, K. V.	27
Montmort, P. R.	17
Newman, C. M.	50, 51, 53
Okner, B.	79
Proschan, F.	51
Radner, R.	1, 2, 7, 8, 12
Ramalingam, T.	2, 40, 53
Randles, R. H.	27, 33
Ranga Rao, R.	60, 65
Rodgers, W. L.	2, 75, 76, 86
Rubin, D. B.	2, 4
Schweizer, B.	49
Serfling, R. J.	45
Shaked, M.	26, 27, 34, 35

Sims, C. A.	13, 79, 80, 81, 86, 89
Srivastava, M. S.	82
Stein, C.	68
Stewart, W. H.	76, 86
Sunter, A. B.	2
Teicher, H.	44, 56
Tong, Y. L.	49
Turner, J. S.	4, 9, 76, 86
Walkup, D. W.	51
Widder, D. V.	35, 63
Wolfe, D. A.	27, 33
Wolff, E. F.	49
Woodbury, M. A.	2, 4, 7
Yahav, J. A.	40, 43, 45, 46, 47
Zionts, S.	90
Zolutikhina	49, 64

Subject Index

Abelian theorem 63
approximate-matching 39, 40
assignment problem 10, 24, 68, 90
associated random variables 53
Base file 6, 69, 75
bias avoiding distance 76, 86, 89
bounded variation 63
broken random sample 14, 18, 19, 20, 21, 22, 24, 39, 56
    re-pairing of 19, 21, 23, 23
complete class of rules 21, 22
completed file 72
concomitants 43
conditional expectation 72
conditionally independent 12, 70, 79, 81, 82, 87, 88, 89, 90
convergence
    almost surely 45
    in factorial moment 40
    in p-th moment 45
    in probability 45, 46
    weak 50
copula 49, 64
correlation 18, 38
distance measures 74
distribution
    bivariate normal 38, 39, 64
    Morgenstern 27, 37, 64, 66
    multivariate normal 22, 68, 70, 71, 78, 81, 82, 86
    Poisson 17, 48, 50, 55, 58
epsilon-correct matching 40
estimation, max, likelihood 81, 83, 84, 85, 87, 89
exchangeable random var. 26, 27, 28, 32, 40, 44, 53
file-merging 1, 14, 70, 79
Framingham Heart Study 2
incomplete files 7
Laplace expansion 48
limiting distribution 47
LPQD property 51, 53
Mahalonobis distance 76, 87, 89
matching 1, 16, 68, 81
    approx. strategies 39, 40
    Bayes strategies 20
    constrained 8, 9, 75
    distance based 70, 75, 78, 86, 89
    epsilon-correct 40
    exact 2, 3, 4
    in bins 89, 91, 92
    isotonic strategy 78, 89
    no-data 17
    packet-level 79

matching
    statistical 2, 4, 6, 7, 70, 78
    unconstrained 8, 9
matching variables 68, 70
max, likelihood pairing 19, 23, 50, 68
micro-data files 5, 13
monotone likelihood ratio 22, 34, 35
monotonicity property 35, 37, 47
Monte-Carlo study 43, 46, 70, 76, 78, 86, 87, 91, 93
number of correct matches 25, 34, 36, 39, 40, 46, 47, 48, 55
optimality criteria 20, 21, 74, 75
optimality properties 25, 34
Poisson limit 47, 48, 49, 50, 55, 56, 58, 68
positive dependent
    association 51
    by expansion 26, 27, 29, 32, 33, 34, 35, 36, 64
    by mixture 26, 27, 28, 29, 32, 33, 34, 35, 36
    conditionally 35, 36, 88
    linear +ve quadrant 51, 53
    positive quadrant 50, 51, 53, 54, 56
PQD property 51, 53, 56
record-linkage 1, 2, 3, 4
supplementary file 6, 8, 71, 75
synthetic file 4, 8, 11, 12, 71, 75
    quality of 13, 25
    reliability of 11, 12, 24, 85
transportation problem 10